SILVIA BORN

Mathematics and Visualization

Series Editors

Gerald Farin
Hans-Christian Hege
David Hoffman
Christopher R. Johnson
Konrad Polthier
Martin Rumpf

Hans-Christian Hege
Konrad Polthier
Gerik Scheuermann
Editors

Topology-Based Methods in Visualization II

With 89 Figures, 78 in Color and 10 Tables

Editors

Hans-Christian Hege
Konrad-Zuse-Zentrum für Informationstechnik Berlin
Devision Scientific Computing
Department Visualization and Data Analysis
Takustraße 7
14195 Berlin
Germany
hege@zib.de

Konrad Polthier
Freie Universität Berlin
Institut für Mathematik
Arnimallee 6
14195 Berlin
Germany
polthier@mi.fu-berlin.de

Gerik Scheuermann
Universität Leipzig
Fakultät für Mathematik und Informatik
Institut für Informatik
Postfach 100920
04009 Leipzig
Germany
scheuermann@informatik.uni-leipzig.de

ISBN 978-3-540-88605-1   e-ISBN 978-3-540-88606-8

Mathematics and Visualization   ISSN 1612-3786

Library of Congress Control Number: 2008938960

Mathematics Subject Classification (2001): 37-04, 37-E35, 37E45, 68U05

© 2009 Springer-Verlag Berlin Heidelberg

This work is subject to copyright. All rights are reserved, whether the whole or part of the material is concerned, specifically the rights of translation, reprinting, reuse of illustrations, recitation, broadcasting, reproduction on microfilm or in any other way, and storage in data banks. Duplication of this publication or parts thereof is permitted only under the provisions of the German Copyright Law of September 9, 1965, in its current version, and permission for use must always be obtained from Springer. Violations are liable to prosecution under the German Copyright Law.

The use of general descriptive names, registered names, trademarks, etc. in this publication does not imply, even in the absence of a specific statement, that such names are exempt from the relevant protective laws and regulations and therefore free for general use.

Cover design: deblik, Berlin

Cover picture by Nelson Max and Tino Weinkauf

Printed on acid-free paper

9 8 7 6 5 4 3 2 1

springer.com

Preface

Visualization research aims to provide insight into large, complicated data sets and the phenomena behind them. While there are different methods of reaching this goal, topological methods stand out for their solid mathematical foundation, which guides the algorithmic analysis and its presentation. Topology-based methods in visualization have been around since the beginning of visualization as a scientific discipline, but they initially played only a minor role. In recent years, interest in topology-based visualization has grown and significant innovation has led to new concepts and successful applications. The latest trends adapt basic topological concepts to precisely express user interests in topological properties of the data.

This book is the outcome of the second workshop on *Topological Methods in Visualization*, which was held March 4–6, 2007 in Kloster Nimbschen near Leipzig, Germany. The workshop brought together more than 40 international researchers to present and discuss the state of the art and new trends in the field of topology-based visualization. Two inspiring invited talks by George Haller, MIT, and Nelson Max, LLNL, were accompanied by 14 presentations by participants and two panel discussions on current and future trends in visualization research.

This book contains thirteen research papers that have been peer-reviewed in a two-stage review process. In the first phase, submitted papers where peer-reviewed by the international program committee. After the workshop accepted papers went through a revision and a second review process taking into account comments from the first round and discussions at the workshop.

About half the papers concern topology-based analysis and visualization of fluid flow simulations; two papers concern more general topological algorithms, while the remaining papers discuss topology-based visualization methods in application areas like biology, medical imaging and electromagnetism.

The book starts with two articles demonstrating the use of finite-time Lyapunov exponents (FTLE) in the visualization of fluid flow simulations (Garth et al., Sadlo and Peikert). The third paper focuses on the calculation of separation surfaces in realistic CFD simulations (Wiebel et al.). It is followed by a paper on topology-based support for the visual analysis of complicated molecules in biology using isosurfaces (Bajaj et al.). The calculation of contour trees for scalar fields on arbitrary meshes is shown by Carr and Snoeyink. Pathline attributes are suggested as an extension of well-known topological concepts to unsteady fields by Shi et al. Salzbrunn and Scheuermann use flow structures based on streamline predicates to select representative streamlines in three- dimensional flows. Max and Weinkauf present a method that is guaranteed to find all critical points of a field generated by a finite set of point charges and they demonstrate its use in the study of electrical fields around molecules. Since the study of chaotic dynamical systems is a real challenge, Krauskopf et al. present a robust algorithm for the calculation of global manifolds; furthermore they demonstrate it, applying it to the well-known Lorenz system. Three applications of topological methods to fluid flow problems follow: first, a topology-guided analysis of vortex breakdown including an approximate analytical model is given (Rütten and Böhme). Second, an article by Peikert and Sadlo studies vortex rings using Poincaré sections. The third article, authored by Laramee et al., identifies several tasks for topological methods in industrial computational fluid dynamics (CFD) analysis based on specific examples. Finally, the contribution by Thomas Wischgoll shows the use of vector field topology for the calculation of center lines in medical imaging data.

Overall, the book presents an informative overview of current research in topology-based visualization and provides insight into various specific research topics.

Acknowledgments

TopoInVis 2007 was organized by the Computer Science Institute of the University of Leipzig, the Zuse Institute Berlin (ZIB) and the Institute of Mathematics of the Freie Universität Berlin. We greatly acknowledge support from the DFG Forschungszentrum MATHEON. We thank all contributors of this volume for the careful preparation of their manuscripts. Special thanks go to Karin Wenzel, Monika Meiler and Alexander Wiebel for their work behind the scenes and help in the production process of this volume.

We are looking forward to the next TopoInVis workshop in February 2009 in Snowbird, Utah.

Berlin and Leipzig,　　　　　　　　　　　　　　　　　　*Hans-Christian Hege*
September 2008　　　　　　　　　　　　　　　　　　　　　*Konrad Polthier*
　　　　　　　　　　　　　　　　　　　　　　　　　　Gerik Scheuermann

Contents

Visualization of Coherent Structures in Transient 2D Flows
*Christoph Garth, Guo-Shi Li, Xavier Tricoche, Charles D. Hansen,
and Hans Hagen* 1

Visualizing Lagrangian Coherent Structures and Comparison
to Vector Field Topology
Filip Sadlo, and Ronald Peikert 15

Extraction of Separation Manifolds using Topological
Structures in Flow Cross Sections
Alexander Wiebel, Xavier Tricoche, and Gerik Scheuermann 31

Topology Based Selection and Curation of Level Sets
Chandrajit Bajaj, Andrew Gillette, and Samrat Goswami 45

Representing Interpolant Topology for Contour Tree
Computation
Hamish Carr, and Jack Snoeyink 59

Path Line Attributes - an Information Visualization Approach
to Analyzing the Dynamic Behavior of 3D Time-Dependent
Flow Fields
*Kuangyu Shi, Holger Theisel, Helwig Hauser, Tino Weinkauf, Kresimir
Matkovic, Hans-Christian Hege, and Hans-Peter Seidel* 75

Flow Structure based 3D Streamline Placement
Tobias Salzbrunn and Gerik Scheuermann 89

Critical Points of the Electric Field from a Collection of Point
Charges
Nelson Max and Tino Weinkauf 101

Visualizing global manifolds during the transition to chaos in the Lorenz system
Bernd Krauskopf, Hinke M Osinga, Eusebius J Doedel 115

Streamline and Vortex Line Analysis of the Vortex Breakdown in a Confined Cylinder Flow
Markus Rütten, Gert Böhme 127

Flow Topology Beyond Skeletons: Visualization of Features in Recirculating Flow
Ronald Peikert and Filip Sadlo 145

Bringing Topology-Based Flow Visualization to the Application Domain
Robert S. Laramee, Guoning Chen, Monika Jankun-Kelly, Eugene Zhang, David Thompson 161

Computing Center-Lines: An Application of Vector Field Topology
Thomas Wischgoll 177

Visualization of Coherent Structures in Transient 2D Flows

Christoph Garth[1], Guo-Shi Li[2], Xavier Tricoche[2], Charles D. Hansen[2], and Hans Hagen[1]

[1] University of Kaiserslautern `garth|hagen@rhrk.uni-kl.de`
[2] University of Utah `lig|tricoche|hansen@sci.utah.edu`

Summary. The depiction of a time-dependent flow in a way that effectively supports the structural analysis of its salient patterns is still a challenging problem for flow visualization research. While a variety of powerful approaches have been investigated for over a decade now, none of them so far has been able to yield representations that effectively combine good visual quality and a physical interpretation that is both intuitive and reliable. Yet, with the huge amount of flow data generated by numerical computations of growing size and complexity, scientists and engineers are faced with a daunting analysis task in which the ability to identify, extract, and display the most meaningful information contained in the data is becoming absolutely indispensable.

1 Introduction

Arguably the major hurdle that hampers the effort of visualization researchers in the post-processing of transient flows is the difficulty to identify proper defining criteria for the coherency of the structures that these flows exhibit. Eulerian approaches focus on the patterns exhibited by streamlines at each instant of time and lend themselves to a topological classification of the flow features. While this leads to visualization algorithms that are computationally efficient and benefit from a strong theoretical framework, the connection of the corresponding structures to the physics of the flow remains unclear. The Lagrangian perspective on the other hand offers a more intuitive account of the material advection induced by the flow but, except in very specific cases, there is an ambiguity attached to the definition of meaningful structures in that setting. For this reason, some visualization techniques, most prominently texture-based approaches, have proposed a variety of ad hoc combinations of Eulerian and Lagrangian perspectives in order to overcome the challenge posed by the ambiguity of patterns that is both coherent in space and time.

In this paper we leverage a concept called *Finite-Time Lyapunov Exponent* (FTLE) that has its roots in dynamical systems theory and has been recently introduced in the fluid dynamics community to resolve this ambiguity. To that end we propose to combine the visual effectiveness of texture-based representations with the physically intuitive meaning of the coherent Lagrangian structures characterized by FTLE. Our method leverages the performance of the Graphics Processing Unit (GPU) to accelerate pre-computation of FTLE and to create expressive animations of the flow that the user can interactively adjust to fit the needs of his visual analysis. We present the application of this approach to the visualization of three different transient flows obtained through Direct Navier-Stokes simulations. While the use of a GPU implementation results in a significant speed-up of the FTLE computation, interactive speeds are still out of reach. For this reason, and due to limitation of space, we wish to concentrate on visualization aspects in the following and do not discuss the technical details of a GPU implementation.

The paper is structured as follows. We first provide a brief introduction to the notion of FTLE. We then discuss some related work in the fluid dynamics and the visualization literature. Section 4 describes and justifies the visualization methods we use while section 5 shows the results obtained for three Computational Fluid Dynamics (CFD) data sets. Finally, we conclude our presentation with a discussion of the benefits and current limitations of our method and we point out interesting avenues for future work.

2 The Finite-Time Lyapunov Exponent

The *finite-time Lyapunov exponent* (FTLE) is a geometric tool that can be used to define and extract coherent structures in transient flows studied in a Lagrangian framework. It has been the object of a growing interest in fluid dynamics research over the last few years and has been successfully applied to a variety of fluid dynamics problems. The Lyapunov exponent is in fact a basic theoretical notion used in the analysis of dynamical systems where it permits to characterize the rate of separation of infinitesimally close trajectories. Its application to aperiodic time-dependent flows, however, has been only recently proposed by Haller [6]. We introduce in the following the basic concepts that are necessary to understand the steps involved in the FTLE computation as we apply them in section 5. As such our presentation is voluntarily informal and we refer the interested reader to the publications listed in section 3 for a more in-depth treatment of this rich subject.

We start by introducing some notations. We consider a time-dependent two-dimensional vector field $\mathbf{v}$ defined over a finite Euclidean domain $U \subset \mathbb{R}^2$ and a (typically finite) temporal domain $I \subset \mathbb{R}$. The position $\mathbf{x}$ of a particle starting at position $\mathbf{x_0}$ at time t_0 after advection along the resulting flow is therefore a function $\mathbf{x}(t, t_0, \mathbf{x_0})$ satisfying $\mathbf{x}(t_0, t_0, \mathbf{x_0}) = \mathbf{x_0}$ and $\frac{\partial \mathbf{x}}{\partial t}\big|_\tau = \mathbf{v}(\tau, \mathbf{x})$. The basic idea behind the notion of FTLE is to define asymptotically stable and unstable coherent structures in terms of loci of maximized dispersion of closely seeded particles. Specifically, consider a fixed initial time

t_0 and a fixed time interval τ, defining $t = t_0 + \tau$. A linearization of the local variations of the map $\mathbf{x}(t, t_0, .)$ around the seed position $\mathbf{x_0}$ is obtained by considering its spatial gradient $J_\mathbf{x}(t, t_0, \mathbf{x_0}) := \nabla_{\mathbf{x_0}} \mathbf{x}(t, t_0, \mathbf{x_0})$ at $\mathbf{x_0}$. We can now use this gradient to determine the dispersion after time τ of particle seeded around $\mathbf{x_0}$ at time t_0 as a function of the direction $\mathbf{d_{t_0}}$ along which we move away from $\mathbf{x_0}$ at t_0: $\mathbf{d_t} = J_\mathbf{x}(t, t_0, \mathbf{x_0}) \mathbf{d_{t_0}}$. Maximizing the norm $|\mathbf{d_t}|$ over all possible unit vector directions $\mathbf{d_{t_0}}$ corresponds to computing the norm of $J_\mathbf{x}(t, t_0, \mathbf{x_0})$ according to the matrix norm $||\mathbf{A}|| := \max_{|\mathbf{x}|=1} |\mathbf{A}\mathbf{x}|$. This norm is known to be the square root of the maximal eigenvalue $\lambda_{\max}$ of the positive definite matrix $\mathbf{A}^T \mathbf{A}$. Therefore maximizing the dispersion of particles around $\mathbf{x_0}$ at t_0 over the space of possible directions around $\mathbf{x_0}$ is equivalent to computing $\sqrt{\lambda_{\max}(J_\mathbf{x}(t, t_0, \mathbf{x_0})^T J_\mathbf{x}(t, t_0, \mathbf{x_0}))}$. This quantity is directly related to the *largest finite-time Lyapunov exponent* $\Lambda(t, t_0, \mathbf{x_0}) = \log(\lambda_{\max}(J_\mathbf{x}(t, t_0, \mathbf{x_0})^T J_\mathbf{x}(t, t_0, \mathbf{x_0}))^{\frac{1}{2}(t-t_0)})$.

Practically, this quantity can be evaluated for both forward and backward advection. Large FTLE values for forward advection correspond to repelling material lines while large FTLE values for backward advection correspond to attracting material lines. Assuming that the set of seed points correspond to the vertices of a grid (e.g. the computational grid), the map $\mathbf{x}(t, t_0, .)$ can be evaluated by numerical integration of pathlines along the flow and its spatial gradient can then be computed with respect to the underlying seeding grid. As noted in [6] the proper identification of attracting and repelling material lines requires to extract ridges from the FTLE field. Ridges of a scalar field α correspond to loci where $\nabla \alpha$ is orthogonal to the minor eigenvector of the Hessian matrix $\nabla^2 \alpha$, under the assumption that the corresponding minor eigenvalue is negative [1]. Observe that the solution proposed in [12] based on the integration of particles along the gradient field of FTLE constitutes in fact an approximation of an actual ridge line computation that is prone to errors. Moreover it has performed poorly in our test cases, due in part to the noise inherently present in our estimates of the FTLE gradient. For these reasons we chose to present in our results the values of FTLE without extracting the corresponding ridges. We show in section 5 that a proper color map is able to emphasize those ridges without explicit extraction of their geometry.

3 Previous Work

As we mentioned previously, Haller has pioneered the use of FTLE as a means to characterize coherent Lagrangian structures in transient flows [6]. In his seminal paper he presented this approach as a geometric one, in contrast to another analytic criterion that he proposed simultaneously based on the notion of preservation of a certain stability type of the velocity gradient along the path of a particle. This work followed previous papers by the same author investigating similar criteria derived from the eigenvectors of the Jacobian of the flow velocity along pathlines to determine the location of Lagrangian coherent structures in the two-dimensional setting [3, 4].

This initial research has generated in the fluid dynamics community a significant interest in FTLE and its applications to the structural analysis of transient flows, both from a theoretical and from a practical viewpoint. Haller proposed a study of the robustness of the coherent structures characterized by FTLE under approximation errors in the velocity field [7]. In the same paper, he suggests to identify attracting and repelling material lines with ridge lines of the FTLE field. Shadden et al. provided a formal discussion of the theory of FTLE and Lagrangian coherent structure [13]. One major contribution of their paper was to offer an estimate of the flow across the ridge lines of FTLE and to show that it is small and typically negligible. An extension of FTLE to arbitrary dimensions is discussed in [10]. These tools have been applied to the study of turbulent flows [5, 2, 12]. They were used in the analysis of vortex ring flows [14]. These notions were also applied to a control problem [8].

On the visualization side of things, multiple approaches have been explored to permit the extraction and the effective depiction of the structures exhibited by time-dependent flows. Topological methods have been applied to transient flows in the Eulerian perspective [18, 16, 17]. Theisel et al. also proposed a method to characterize the structure of pathlines by subdividing the domain into sink, source, and saddle-like regions based on the divergence of the restriction of the flow to a plane orthogonal to the pathline orientation in space-time [17].

Additionally, texture-based representations have been considered to visualize time-dependent flows while offering an effective depiction of salient structures, see [9] and references therein. Because of the intrinsic difficulty of defining structures that are both coherent in space and time, each of these methods resorts to some form of ad hoc way to combine the Eulerian and Lagrangian perspectives, leading to animations for which a physical interpretation is typically ambiguous.

In the present work we therefore propose to combine a texture-based representation method called *GPUFLIC* that we introduced recently [11] with a visual encoding of FTLE in order to emphasize meaningful patterns in a common flow visualization modality and to clarify their relationship with coherent Lagrangian structures.

4 Visualization of Coherent Structures

In this section, we show how a direct visualization of the FTLE field for a given flow can be achieved. While the direct numerical computation of the FTLE for a dense sampling of a given flow region with adequate resolution is usually prohibitively expensive, we were able to reduce computation times significantly be employing the computational power available through the use of commodity graphics hardware (GPU). We will not give our method here, due to space considerations, but will present it in forthcoming work. Instead, we will focus on visualization of the results of this computation.

4.1 Direct FTLE Visualization

The earliest work on direct FTLE visualization was again done by Haller [6], who used a dense color mapping to visualize basic FTLE structures that covers all primary colors. With this approach, Lagrangian coherent structures appear as local maximizing lines of the FTLE field. However, his visualization is unfortunate in the sense that maximizing lines are not intuitively identifiable with a single color. If weaker coherent structures exist, the may have a different color than the stronger structures elsewhere in the field. Therefore, this technique does not lend itself well to an intuitive understanding. One possible remedy for this is a ridge extraction followed by the visualization of these locally maximizing lines. However, these approaches are usually highly sensitive to numerical issues, and can result in false positives. Moreover, coherent structures are presented in a skeletonized fashion, clearly describing their existence but not the relative strength.

A second topic of importance is the temporal orientation of the FTLE computation. Looking at the FTLE in forward time, it is essentially a measure of the maximal stretching of pathlines, and is therefore a good candidate to visualize coherent structures of a diverging nature. To achieve similar results for converging structures, it is necessary to also look at the FTLE in backward time, i.e. compute the measure of pathline convergence. In the following, we will abbreviate these two different scalar measures $FTLE^+$ and $FTLE^-$ to indicate forward resp. backward temporal orientation. Again, naive color mapping has difficulties of representing these two quantities for incompressible flows, since they often show regions of saddle-type behavior where both $FTLE^+$ and $FTLE^-$ have a significant value together, simply because they do not follow an exclusive-or relationship.

We therefore propose a two-dimensional color mapping scheme. First, we normalize the FTLE fields over the space-time domain of the given dataset to the unit interval, using the same normalization on both fields in order to preserve the relative strength of coherent structures. Then, we apply the two-dimensional colormap presented in Fig. 4, resulting in a visualization that represents converging structures in blue and diverging structures in red, encoding the relative strength through saturation. Figure 1 provides an example. We will discuss how this enables an analysis of typical flow patterns from FTLE visualizations in more detail in Section 5.

Although geometric context can be provided through explicit depiction of domain boundaries and objects embedded in the flow, a more flow-centric context is often needed to interpret FTLE visualizations in terms of flow mechanics. In the next section, we will briefly discuss a simple yet intuitive approach which, in combination with direct FTLE visualization, can provide insightful visualizations.

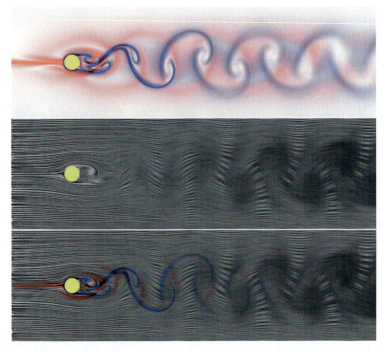

Fig. 1. Comparison: Direct FTLE visualization **(top)**, UFLIC **(middle)**, a combination of both **(bottom)**

4.2 GPUFLIC

GPUFLIC [11] is the texture-based flow visualization method that we use in conjunction with FTLE to depict the evolution of salient structures in the transient flows introduced in the next section. This method constitutes an efficient implementation on the GPU of an algorithm proposed by Shen and Kao called UFLIC [15]. The basic idea of this scheme consists in advecting a dense set of particles that deposit the color attribute that they carry as they traverse the space time domain. At each instant in time, each pixel of the texture covering the domain averages the contributions made by all the particles that crossed it during the last time step, which creates a frame of the animation. Additionally, to maintain a good contrast as the animation progresses, a high-pass filtering step is applied to each frame and combined with an input noise to produce the color attributes assigned to a whole new set of particles – one per pixel – that are injected into the flow at each time step. Refer to [11] for a more detailed description of the algorithm.

4.3 In-context FTLE Visualization

To provide the necessary context for FTLE visualization, we propose a simple combination of the FTLE and GPUFLIC flow visualizations. By multiplicatively weighting the color channels in the images generated by both methods for the same domain (with the same integration parameters), we obtain more expressive visualizations (see e.g. Fig. 1). This is essentially facilitated by the dense nature of GPUFLIC in contrast to the sparse representation typically obtained with FTLE in the presence of clearly defined structures. Where GPU-FLIC expresses the basic Lagrangian information such as flow direction and magnitude, FTLE complements this nicely with information about coherence and convergence.

A slightly different visualization effect is achieved by encoding FTLE information into a single scalar field, based on the balance of convergence and divergence of pathlines. Since, in the total absence of either, the flow is uniform, these properties can be augmented with an interpretation in terms of forcing the flow into certain patterns: at a point of high divergence, the flow is forced away from this point. The opposite is true for points of high convergence. A good analogy is the gradient field of a scalar height field. Therefore, by encoding the balance between $FTLE^+$ and $FTLE^-$ in a scalar field, we draw on a natural physical understanding of height fields and their gradients. Again, the resulting visualization is enhanced with the FTLE and GPUFLIC results as texture. Fig. 6 demonstrates this.

Having discussed several possible techniques to apply the FTLE for direct visualization of two-dimensional flow fields, we move on to specific datasets and the visualization results obtained there using these methods.

5 Results

All specific visualization examples described in this section center on results obtained from CFD datasets. They are adaptive-resolution, time-adaptive unsteady direct numerical simulations of the incompressible Navier-Stokes equations. Three basic types of flow were chosen because each of them lends itself well to illustrate different aspects of FTLE-based visualization. We only treat the first dataset in detail and, due to space limitations, only present basic results for the other two.

5.1 Kármán Vortex Street

The Kármán Vortex Street is one of the most widely known patterns in fluid mechanics. It consists of a vortex street behind a cylinder and is a special case of unsteady flow separation from bluff bodies embedded in the flow. It is quite well understood and therefore an ideal test case for many applications.

Figure 1 illustrates the basic modes of direct unsteady flow visualization presented in this work. The top image shows only the direct FTLE color map. It identifies the separation structures behind the cylinder (red) that separate the region of vortex genesis directly behind the cylinder from the surrounding flow. The curved attachment structures (blue) visualize the convergence of material at the vortices. As the flow moves away from the cylinder, these structures are essentially advected and grow weaker. We observe that for a ridge-line type visualization (without encoding of feature strength), the weakening of structures would not be observable. The combination of FTLE and GPUFLIC (bottom image of Fig. 1) allows for a visual identification of individual vortical structures. Since they are necessarily counterrotating, FTLE depicts the boundaries as line-type divergent regions. GPUFLIC alone does not achieve an identification of structures except close to the cylinder (middle image) and is only comprehensible if enhanced by FTLE visualization (Fig. 6).

We further employ this well-understood example to study some of the properties of the Finite-Time Lyapunov Exponent. Figure 5 shows the effect of different integration lengths on the FTLE computation. As a general rule, coherent structures become more pronounced with increasing integration time, as pointed out in [12]. On the other hand, long integration (in comparison to the reference time, in other words the natural time scale of the problem) may yield coherent structures that are not actually meaningful for short-term flow evolution, see [7]. We conclude from this that the reference time, which can be interpreted as a measure for the rate of change of the flow field, seems like a good choice for the integration length.

A related topic is the application of FTLE visualization to stationary flow fields. If the integration length is (theoretically) increased to infinity in such fields, the local maximum lines of the resulting coherent structures should coincide with the topological graph of the flow field. Figure 2 illustrates the resulting features (top image) in comparison to the unsteady results (bottom image). Behind the cylinder, the structures are very topological in nature, i.e. the are attracting and repelling material lines intersecting at saddle-like points. However, the vortical characterization is lost as the flow is advected (topology is not Galilean invariant), and the flow pattern is unclear. Overall, these results suggest that FTLE-type analysis is a possible adaptation of topological methods to unsteady flows.

5.2 Heated Cylinder Flow

This example was computed using the classical Boussinesq approximation to simulate the flow generated by a heated cylinder. This approximation adds a source term proportional to the temperature (modeled as a diffusive material property) to the vertical component of the velocity field. The cylinder serves as a temperature source and thereby generates a *plume* of upward flowing

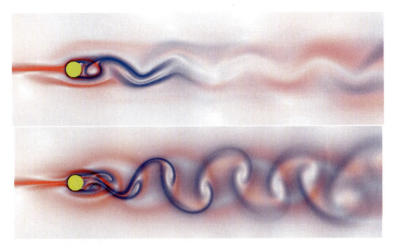

Fig. 2. Comparison: Direct FTLE visualization, steady case **(top)**, unsteady case **(bottom)**

material. As the plume moves upward, its outer layers exchange heat with the surrounding flow, resulting in inhomogeneous friction and hence turbulent flow. Figure 7 shows four timesteps of the resulting visualization using the proposed methods. The interpretation here is more difficult than for the previous dataset since there is more structure on smaller scales, and hence more detail. Quite evident, however, is the clear separation of plume-related flow from the overall surrounding flow.

5.3 Rayleigh-Taylor instability

The term "Rayleigh-Taylor instability" essentially refers to the interactions of two fluids of different density. In this example, we have chosen an initial configuration in which a denser fluid rests on top of a lighter fluid. As the simulation progresses, the upper fluid is drawn downwards by gravity, resulting in the typical (inverse) mushroom structure. This flow differs from the already presented examples in several respects.

First, the time resolution is limited by the appearance of small-scale structures at the material interface whose temporal evolution must be correctly resolved. The resulting GPUFLIC visualization is unsatisfying due to the strong variation of scales that makes it difficult to choose an integration time that will equally accentuate all relevant structures. Therefore, to provide context to the FTLE visualization, we have added the material interface to the resulting images (Figure 3, black lines). This provides enough orientation to make results comprehensible.

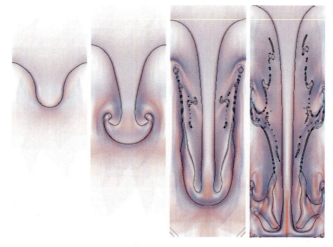

Fig. 3. Direct FTLE visualization for the Rayleigh-Taylor instability, context is provided by the boundary between the two fluids

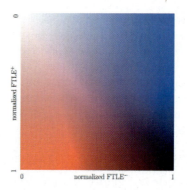

Fig. 4. Colormap

The second difference to the previous examples is the relative weakness and large extent of coherent structures. While one would expect the material interface to show up clearly in the FTLE images, this is not the case. Enlarging the integration time to provide for more coherent structures is ruled out by the range of the simulation, since clearly it is not possible to integrate pathlines past the end of the simulation. We conclude that this flow contains only few long-term coherent structures and is mainly driven by small-scale motion. In comparison to the previous examples, FTLE values are an order of magnitude smaller, which is obscured in visualization by the normalization we apply (see 4).

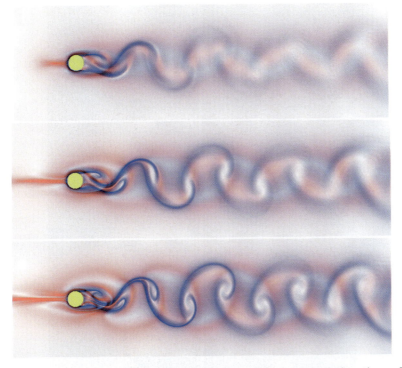

Fig. 5. Comparison: Direct FTLE visualization, different integration times. 0.25 (**top**), 0.5 (**middle**), 0.75 (**bottom**)

Fig. 6. Height field visualization of FTLE with UFLIC texture and FTLE color coding

6 Discussion

In this paper, we have empirically studied the Finite Time Lyapunov Exponent and its applications in the visualization of time-dependent planar flows. We have shown how the visualization can be greatly enhanced by an explicit choice of color mapping and a combination with the GPUFLIC technique. Furthermore, we have taken first steps to examine various aspects of the FTLE such as

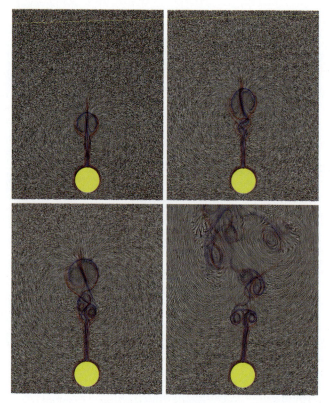

Fig. 7. Four timesteps of the heated cylinder flow, with evolution from top-left to bottom-right

dependence on integration time and application to steady flows. These aspects were illustrated on three typical examples of unsteady, two-dimensional flows.

Future work seems promising and manifold: While we were able to produce good visualization results for the test cases, we would like to study application examples, where the FTLE can possibly help in solving some of the more difficult problems in flow analysis, like the extraction of separation and attachment lines on curve surfaces. As a general theme, a generalization of the presented concepts and algorithms to higher dimensions seems necessary to study unsteady volume flows.

Acknowledgements

Parts of this work were financially support by DFG and the International Research Training Group (IRTG) at the University of Kaiserslautern.

We would especially like to thank Gordon Kindlmann for making his TEEM software publicly available. It has proven very helpful in the process of developing the visualizations presented here.

References

1. D. Eberly, R. Gardner, B. Morse, and S. Pizer. Ridges for image analysis. *Journal of Mathematical Imaging and Vision*, 4:355–371, 1994.
2. M.A. Green, C.W. Rowley, and G. Haller. Detection of lagrangian coherent structures in 3d turbulence. *J. Fluid Mech.*, to appear, 2006.
3. G. Haller. Finding finite-time invariant manifolds in two-dimensional velocity fields. *Chaos*, 10(1):99–108, 2000.
4. G. Haller and G. Yuan. Lagrangian coherent structures and mixing in two-dimensional turbulence. *Physica D*, 147:352–370, 2000.
5. G. Haller. Lagrangian structures and the rate of strain in a partition of two-dimensional turbulence. *Physics of Fluids*, 13(11), 2001.
6. G. Haller. Distinguished material surfaces and coherent structures in three-dimensional flows. *Physica D*, 149:248–277, 2001.
7. G. Haller. Lagrangian coherent structures from approximate velocity data. *Physics of Fluids*, 14(6):1851–1861, june 2002.
8. T. Inanc, S.C. Shadden, and J.E. Marsden. Optimal trajectory generation in ocean flows. In *Proceedings of the American Control Conference*, pages 674–679, 2005.
9. B. Laramee, J.J. van Wijk, B. Jobard, and H. Hauser. ISA and IBFVS: Image space based visualization of flow on surfaces. *IEEE Transactions on Visualization and Computer Graphics*, 10(6):637–648, nov 2004.
10. F. Lekien, S.C. Shadden, and J.E. Marsden. Lagrangian coherent structures in n-dimensional systems. *Physica D*, submitted, 2006.
11. G.-S. Li, X. Tricoche, and C.D. Hansen. GPUFLIC: Interactive and dense visualization of unsteady flows. In *Data Analysis 2006: Proceedings of Joint IEEE VGTC and EG Symposium on Visualization (EuroVis) 2006*, pages 29–34, may 2006.
12. M. Mathur, G. Haller, T. Peacock, J.E. Ruppert-Felsot, and H.L. Swinney. Uncovering the lagrangian skeleton of turbulence. *Phys. Rev. Lett.*, submitted, 2006.
13. S.C. Shadden, F. Lekien, and J.E. Marsden. Definition and properties of lagrangian coherent structures from finit-time lyapunov exponents in two-dimensional aperiodic flows. *Physica D*, 212:271–304, 2005.
14. S.C. Shadden, J.O. Dabiri, and J.E. Marsden. Lagrangian analysis of fluid transport in empirical vortex ring flows. *Physics of Fluids*, 18:047105, 2006.
15. H.W. Shen and D.L. Kao. A new line integral convolution algorithm for visualizing unsteady flows. *IEEE Transactions on Visualization and Computer Graphics*, 4(2), 1998.
16. H. Theisel and H.-P. Seidel. Feature flow fields. In *Proceedings of Joint Eurographics - IEEE TCVG Symposium on Visualization (VisSym '03)*, pages 141–148, 2003.
17. H. Theisel, T. Weinkauf, H.-C. Hege, and H.-P. Seidel. Topological methods for 2d time-dependent vector fields based on streamlines and path lines. *IEEE Transations on Visualization and Computer Graphics*, 11(4):383–394, 2005.
18. X. Tricoche, T. Wischgoll, G. Scheuermannn, and H. Hagen. Topology tracking for the visualization of time-dependent two-dimensional flows. *Computer and Graphics*, 26:249–257, 2002.

Visualizing Lagrangian Coherent Structures and Comparison to Vector Field Topology

Filip Sadlo and Ronald Peikert

Computer Graphics Laboratory, Computer Science Department,
ETH Zurich, Switzerland {sadlo, peikert}@inf.ethz.ch

Summary. This paper takes a look at the visualization side of vector field analysis based on Lagrangian coherent structures. The Lagrangian coherent structures are extracted as height ridges of finite-time Lyapunov exponent fields. The resulting visualizations are compared to those from traditional instantaneous vector field topology of steady and unsteady vector fields: they often provide more and better interpretable information. The examination is applied to 3D vector fields from a dynamical system and practical CFD simulations.

1 Introduction

Vector field topology (VFT) is often used to obtain a simplified representation of a vector field or phase space of a dynamical system. Introduced to the visualization community by Helman et al. [7], it also allows deeper insight into the structure of vector fields. VFT deals with the detection, classification and global analysis of critical points (isolated zeros of the vector field). The manifolds that are defined by the eigenvectors of the velocity gradient at these points can be computed by integrating streamlines (for 1D manifolds) or stream surfaces (for 2D manifolds). According to the eigenvalues, the manifolds can be stable (negative real part) or unstable (positive real part). In other words, a stable manifold is the set of all trajectories that converge to the critical point in positive time [1]. The manifolds are also called separatrices because they separate regions of different flow behavior in the respective direction of time.

However, there is one important drawback of the method: it is meaningful in a direct sense only for steady vector fields (autonomous dynamical systems). One reason for this limitation is that pathlines usually diverge from streamlines and that critical points often move in unsteady vector fields. Unsteady vector fields are often analyzed by VFT of isolated time steps. Although this is hard to interpret and gives no precise information about the true behavior, it gives an instantaneous picture and can give insight especially when applied to derived fields. Other approaches to a time-dependent topology based on path lines are that of Theisel et al. [16] and Shi et al. [14].

The advantage of the concept of coherent structures (Section 2) is that it shows the true behavior, is clearly physically motivated, scale-aware and therefore noise-insensitive, and easy interpretable, even for unsteady vector fields.

This paper describes the concept of Lagrangian coherent structures and how they are obtained by filtered ridge extraction from finite-time Lyapunov exponent in Section 2. In Section 3 FTLE ridges are extracted from steady and unsteady 3D vector field examples and compared to vector field topology.

2 Lagrangian Coherent Structures

In recent years, the concept of Lagrangian coherent structures (LCS) is attracting attention in the field of vector field analysis, especially since Haller [4] has shown that LCS can be obtained by detecting local extrema in the finite-time Lyapunov exponent (FTLE) (explained in Section 2.1). Material lines or surfaces (LCS) are attracting if infinitesimal perturbations converge to these structures in forward time and repelling if they are attracting in backward time. According to Haller [4], attracting LCS can be obtained as local maxima, or ridges (approximated as height ridges described in Section 2.2), of backward-time FTLE, and the repelling ones as ridges in forward-time FTLE. Stable and unstable manifolds tend to have its analog in repelling and attracting material lines or surfaces, at least for steady vector fields (see also results in Section 3). In contrast to vector field topology, LCS tend to be insensitive to short-term perturbations and small-scale noise, such as turbulence, due to their Lagrangian definition. Additionally, LCS are usually more appropriate for unsteady vector fields due to their clear physical motivation and interpretability. Note that LCS of unsteady fields usually deform and move over time but are still easy to interpret.

2.1 Finite-Time Lyapunov Exponent

The finite-time Lyapunov exponent (FTLE) measures the separation (or expansion) rate of nearby particles when advected by the flow for a given time T. For a n-dimensional vector field, there are n Lyapunov exponents. Here we are only interested in the largest FTLE. It is a scalar Lagrangian measure stored at the starting point of the respective trajectory. According to Haller [4] the FTLE can be computed by advecting each sample point $\mathbf{x} \in D$ of an arbitrary grid at time t_0 with the flow for time T, resulting in a flow map $\phi_{t_0}^{t_0+T}(\mathbf{x})$ that maps $\mathbf{x}$ to its advected position. We decided to stop the advection if the point reaches a domain boundary and store the position on the boundary in the flow map.

The maximum separation of two close particles can be computed from the gradient of the flow map: it is the spectral norm of its gradient. In other words: to measure the maximum separation one has to seed the two particles along the direction of maximum expansion, which is the direction of the eigenvector belonging to the largest eigenvalue of

$$\Delta(\mathbf{x}) = (\nabla \phi_{t_0}^{t_0+T}(\mathbf{x}))^\top \cdot \nabla \phi_{t_0}^{t_0+T}(\mathbf{x}). \tag{1}$$

(1) is called the Right Cauchy-Green deformation tensor, measuring the square of the distance change due to deformation. Accordingly, the maximum separation is the square root of the largest eigenvalue of $\Delta(\mathbf{x})$. Lyapunov exponents are used to measure exponential growth rates of perturbations. Therefore the logarithm of the resulting value is computed and additionally normalized by absolute advection time $|T|$, leading to the following formulation for the largest FTLE denoted as $\sigma_{t_0}^T$:

$$\sigma_{t_0}^T(\mathbf{x}) = \frac{1}{|T|} \ln \sqrt{\lambda_{max}(\Delta(\mathbf{x}))}. \tag{2}$$

The reader is referred to the work of Haller for further information on LCS and FTLE [4, 5], and vortices and FTLE [6].

2.2 Height Ridges

Height ridges are local maxima in a relaxed sense. More precisely, height ridges are locations where a scalar field s has a local maximum in at least one direction. More general, height ridges are d-dimensional manifolds in n-dimensional space with $n > d \geq 0$.

The ridge criterion can be formulated using the gradient and the Hessian of s. Note that for a height ridge, the eigenvectors belonging to the d largest eigenvalues λ_i ($i = 1, \ldots, d$) of the Hessian point along the ridge, whereas the eigenvectors of the $(n - d)$ smallest eigenvalues λ_j ($j = d+1, \ldots, n$) point orthogonally to the ridge. One necessary condition for a ridge is that the derivatives in λ_j-eigenvector directions are zero. This leads to the condition

$$\boldsymbol{\epsilon}_{\lambda_j} \cdot \nabla s = 0 \tag{3}$$

with $\boldsymbol{\epsilon}_{\lambda_j}$ the eigenvector belonging to λ_j. The other condition for a height ridge is that the second derivatives in $\boldsymbol{\epsilon}_{\lambda_j}$ directions are negative, formulated as

$$\lambda_j < 0. \tag{4}$$

Valley lines (the opposite of height ridges) are obtained by computing height ridges of the field $-s$. The reader is referred to the work of Eberly [2], Lindeberg [12], and the thesis of Majer [13] for further details.

For the extraction of 2D ridges in the 3D domain, one would like to use e.g. *Marching Cubes*. However, since an eigenvector is not oriented, direct application of these methods to (3) fails because the eigenvectors at the nodes of a cell can be inconsistently oriented. For their *Marching Ridges*, Furst et al. [3] use PCA to achieve local consistency of the eigenvectors of a cell. Kindlmann et al. [10] achieve per-cell eigenvector consistency by sampling along each edge of the cell and observing eigenvector rotation. Inter-cell consistency is circumvented by a subsequent pass over the triangles that fixes their orientation.

In this work, per-cell eigenvector consistency is guided by PCA of the eigenvectors at the nodes of the cell, according to Furst et al.. We experienced non-orientable ridge surfaces in some applications. Flat shading and bi-directional lighting was chosen in these cases.

2.3 FTLE Ridge Filtering

Because ridge extraction involves computation of second derivatives, noise amplification can become an issue. Smoothing is applied in these cases in order to obtain significant visualizations. One has to keep in mind however, that this tends to deform the LCS, i.e. particles can permeate the computed FTLE ridges to a certain degree. It is therefore advisable to verify the LCS using trajectories (for steady vector fields) or animations of LCS and particles (unsteady vector fields).

Smoothing is realized by incorporating it into the gradient computation. In our case, the gradient at a given node is computed by fitting a linear vector field to its neighboring nodes in a Least Squares sense. The degree of smoothing can be controlled by adjustment of the neighborhood range.

The finite-time Lyapunov exponent measures the amount of separation. It is therefore straight-forward to use it to filter out parts of ridges with low separation property. This approach is physically motivated and therefore results in relevant and consistent visualizations. It is therefore our favorite method for FTLE ridge filtering.

Filtering connected components of the final mesh by their area is also an effective method for improving the visualization. Small connected components of ridges are likely to be noise, as long as the other filtering conditions did not disrupt the ridges because of low tolerance.

Another approach is to use the second derivative across the ridge (λ_n) for filtering out "flat" ridges. Although it turned out that its effect was comparable to filtering by FTLE in our examples, it is only geometrically motivated and therefore less preferable. Therefore it was not used for the results in Section 3, except for the vortex ring in Section 3.4.

In order to filter out ridges that arise due to trajectories that reach the domain boundary, it is allowed to filter out ridge regions by advection time of the corresponding trajectories. As noted in Section 2.1, pathline integration is stopped if the particle reaches a domain boundary. The advection time is smaller than T in these cases and a threshold can be used for suppressing them.

During ridge extraction by marching ridges (Section 2.2), the necessary ridge condition (4) and filtering conditions are tested at the vertices of the resulting triangles and triangles that violate them are rejected. Triangle trimming was not implemented in the current approach, leading to zigzag ridge borders. Figures 9(c)–9(e) show an example of FTLE ridge filtering.

3 Results

The described methods are applied to different vector fields. The first example is the analytic and steady ABC flow (Section 3.1). Then 3D saddles in isolated time steps of an unsteady Francis water turbine CFD simulation are examined (Section 3.2), and vector field topology is compared to FTLE ridges. In Section 3.3 the flow around the divider of the same CFD result is analyzed but this time both, in a steady and unsteady manner. Section 3.4 takes again a look at the Francis dataset, but this time at two vortices. Finally, Section 3.5 examines a steady-type Pelton water turbine CFD simulation.

3.1 ABC Flow

Vector field topology and FTLE ridges are applied to the analytic steady ABC flow field. This flow has three parameters A, B, and C, (in this example set to $\sqrt{3}$, $\sqrt{2}$, and 1 according to Henon [8]) named after the researchers Arnold, Beltrami, and Childress, and can be written as the dynamical system

$$\begin{aligned} \dot{x} &= A \sin z + C \cos y \\ \dot{y} &= B \sin x + A \cos z \\ \dot{z} &= C \sin y + B \cos x \end{aligned} \qquad (5)$$

It is triple-periodic in space and divergence-free. Despite of its simple Eulerian nature, it exhibits complicated Lagrangian structure such as invariant tori and chaotic advection [9] if considered as a three-dimensional torus. Other interesting properties are that it is identical to its curl and therefore fully helical. This is the cause why vortex core line detection based on helicity, such as that by Levy et al. [11], fail on this flow. The ABC field was discretized on a regular grid in order to show the applicability of the method to practical vector fields.

Figure 8(a) shows the VFT view to the field. Critical points have been determined and streamlines have been computed in positive and negative time. Streamlines are seeded on two rings of seeds around the critical point. The circles are usually chosen coplanar with the 2D manifold, have user defined radius and user defined offset along the direction of the 1D manifold. Unfortunately it turns out that the 2D manifolds are degenerate in this case of the ABC flow, meaning that one eigenvalue of the velocity gradient is zero. Therefore this is a steady case where the vector field topology approach fails or may not be practical to give a complete image of the flow structure.

Haller already investigated the ABC flow using FTLE [4], but without extracting ridges. Figures 8(b)–8(d) show the LCS view to the field using ridges. It can be seen that the ridges are consistent with the manifolds: the critical points are located at the intersection of positive-time and negative-time FTLE ridges, and the streamlines in positive and negative time follow the corresponding ridges. FTLE was only computed at the "original" nodes in the first period of the ABC flow consisting of 30^3 nodes, but the integration time of 2 caused the trajectories to reach neighboring periods of the ABC flow as well. Ridges were only generated in regions with FTLE higher or equal to 0.9 for suppressing weak separation phenomena, gradient neighborhood range for smoothing during ridge extraction was 2 (as in all examples), and connected components below 50 triangles have been suppressed. The computation took 113 seconds. Figures 9(a) and 9(b) show another view and some of the trajectories used for computation.

To also visualize the short-time separation aspect of the flow, short-time FTLE was computed and color-coded on the long-time FTLE ridges. Figure 8(e) shows the result of integration time +0.001 (which took 22 seconds) on the positive-time ridges. It can be seen that the short-time FTLE exhibits local maxima near the critical points. Figure 8(f) shows additionally the negative-time ridges with short-time FTLE of integration time −0.001. There are also local maxima of the FTLE near the critical points in negative time direction. From the streamlines it can be seen, that the local maxima are not in upstream or downstream direction of the critical points, as one may assume. It has to be investigated to what extent this situation is sensitive to noise and if it is a specialty of the ABC flow or a general principle.

3.2 3D Saddles in Francis Draft Tube

In this section the LCS and VFT approaches are compared for non-spiralling 3D saddles. As a first step one would think of applying the methods to an analytic linear vector field containing a saddle, described by a Jacobian with real eigenvalues. VFT performs well in these cases, unless the 2D manifold of the saddle is degenerate as in Section 3.1. However, the FTLE ridge approach is not able to capture linear saddles because all trajectories through it would exhibit the same FTLE value and therefore there would be no ridges corresponding to its manifolds. This is a drawback of the FTLE ridge approach.

However, it is unlikely that purely linear saddle regions appear in practical vector fields. Therefore the examination was applied to some of the saddles in a CFD simulation of the draft tube of a Francis water turbine. As a first approach, a single time step of the unsteady simulation was used for the analysis. This results in instantaneous LCS based on streamlines, suited for

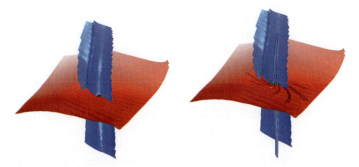

Fig. 1. A 3D saddle in Francis draft tube. Same visualization as in Figures 9(f) and 9(g)

the comparison of VFT and LCS methods. The critical points were detected and at each critical point two FTLE were computed on a regular grid of 60^3 nodes around the critical point, one with integration time $+1$ second and one with -1 second. The choice of the integration time was based on a-priori knowledge of the data. Computation at the first saddle took 321 and 427 seconds, respectively. Confer 9(d) and 9(e) for filtering details. The extracted ridges are shown in Figures 9(f), 9(g), and 1. Figure 4 shows some streamlines used for FTLE computation of Figure 1.

One can see that the 2D manifolds are well captured by the corresponding ridges, resulting in smooth surfaces. It has to be noted that also the opposite-time FTLE ridges result in surfaces, even though these surfaces exhibit more curvature and folding. It can be seen that these ridges are well consistent with the 1D manifolds of the saddles. We conclude that: the examined critical points lie on the intersection curves of positive-time and negative-time FTLE ridges. This was also observed in the ABC flow example of Section 3.1. The 2D manifolds have a ridge counterpart and the 1D manifolds are consistent with the corresponding opposite-time 2D ridge. Therefore, generating FTLE ridges in positive and negative time in regions around critical points tends to convey more information than traditional VFT and can serve as topological icons. Extracting and visualizing the intersection curves of positive-time and negative-time FTLE ridges, similar to the saddle connectors of Theisel et al. [15], seems promising and could serve as a kind of a topological skeleton, which could be applicable even for unsteady vector fields.

Although our investigation did not result in any "purely linear" 3D saddle regions in CFD simulations, it has to be examined how frequent they are in practical vector fields and what extent they have. The extent is of some importance because the FTLE ridge approach fails if the trajectories do not escape from the linear regime of the vector field. Another thing to note is that for short advection times $|T|$ the FTLE ridges tend to be less smooth, smaller, and less consistent with VFT. This turned out to be a problem for getting the unsteady positive and negative time FTLE ridges on the saddles: the temporal domain of the simulation was too short with respect to the low velocities in the region where the saddles reside.

3.3 Bifurcation in Francis Draft Tube

In this section the unsteady CFD flow around the divider of the Francis draft tube is analyzed using steady and unsteady FTLE ridges. The divider is a construct that divides the flow into the two channels. First, instantaneous FTLE ridges were computed at the first time step of the simulation. Figure 2 (left) shows some of the positive-time streamlines used for FTLE computation, Figure 2 (right) shows additionally the resulting ridge. One can see that the ridge is deformed at the horizontal vortex core line (computed according to Levy et al. as in all examples) in the upper part of the image. However, the ridge does not exhibit a hole where it intersects that vortex core line. For the instantaneous flow, this can be interpreted that the flow passes the vortex core line at a critical point and is finally separated at the divider. On the other hand, the ridge forms a tunnel around the vortex core line at the bottom of the image. This is a case where the vortex is captured as a distinct LCS.

Next, the instantaneous FTLE ridge of the first time step is compared to the unsteady FTLE ridge of the first time step. Figure 3 shows the corresponding visualizations. Both steady and unsteady FTLE ridges were computed on a $30 \times 40 \times 50$ grid using an advection time of 0.4 seconds and filtered by requiring a minimum FTLE of 7.1. The computation took 631 seconds in the steady case and 1255 seconds in the unsteady case. In order to remove other

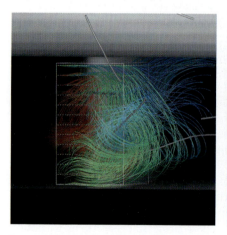 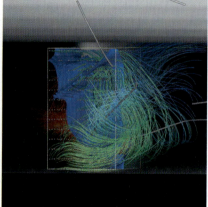

Fig. 2. Flow in Francis draft tube. Left: Vortex core lines (gray tubes) and some streamlines used for FTLE computation (arbitrary colored tubes started at white spheres). Right: Additionally instantaneous positive-time FTLE ridge visualizing the bifurcation at the divider

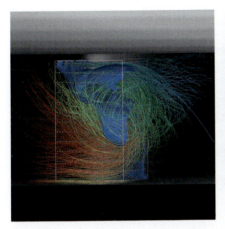 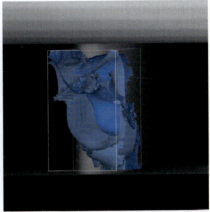

Fig. 3. Flow in Francis draft tube. Left: Unsteady positive-time FTLE ridge (blue) visualizing the bifurcation at the divider with some path lines used for FTLE computation (arbitrary colored tubes). Vortex core lines have been omitted because they move in time. Right: Comparing instantaneous positive-time FTLE ridge (light blue) with unsteady FTLE ridge

Fig. 4. Opposite view to the visualization of Figure 1 (left), with arbitrary-color tubes visualizing some of the positive-time streamlines

ridges that were not consistent with the ridge under consideration, the minimal connected component size was set to 2000 triangles. It is clearly visible that the unsteady FTLE ridge differs in shape from the steady FTLE ridge. One difference is that it does not divide the flow on the left hand side anymore. Instead, it extends only to the right. Some trajectories are crossing the unsteady ridge. This is likely to happen for unsteady LCS because they are material surfaces at a given time whereas trajectories extend over time

Fig. 5. Steady positive-time FTLE ridges (blue) around vortex core line (gray tube) in front of the Francis divider

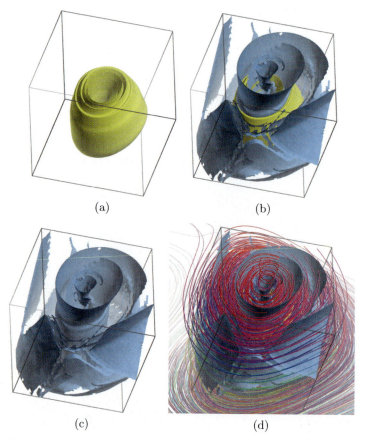

Fig. 6. Vortex breakdown bubble in Francis draft tube. (a) Unstable manifold of vortex breakdown bubble. (b) Same as (a) with instantaneous negative-time FTLE ridges. Ridge is consistent with fold. (c) Ridge from (b): cylindrical inside bubble. (d) Same as (b) with some negative-time streamlines used for FTLE computation

and are immaterial. Trajectories with nearby seeds are visualizing the mechanisms of separation.

3.4 Vortices in Francis Draft Tube

Vortices are coherent structures and therefore they should show up in FTLE ridge visualizations. Two vortices are examined, both using instantaneous FTLE ridges because of the small temporal domain of the underlying CFD simulation. The first vortex is in front of the divider from Section 3.3, but this time at the last time step. Figure 5 shows the positive-time FTLE ridges of this vortex. The vortex is nicely captured by the FTLE ridge that is also indicating the separation by the divider. The grid consisted of $30 \times 40 \times 50$ nodes, advection time was 0.4 seconds, ridge regions with FTLE smaller than 5.5 were suppressed, as well as connected components smaller than 1000 triangles, and the computation took 774 seconds.

The second one is a vortex ring (vortex breakdown bubble) in the right channel of the draft tube. Figure 6 shows its unstable manifold and negative-time FTLE ridges. Interestingly, the corresponding ridge does not exhibit the bubble shape of the manifold, it is simply cylindrical, although consistent with the fold of the manifold. FTLE was computed on a 60^3 grid with 4 seconds advection time, which took 704 seconds to compute. Ridge regions with $\lambda_n < 300$ were suppressed in order to remove noise, as well as connected components smaller than 2000 triangles.

3.5 Bifurcation in Pelton Distributor Ring

In this section, the steady CFD flow inside the distributor ring of a Pelton water turbine is examined using FTLE ridges. Figure 7 shows positive-time

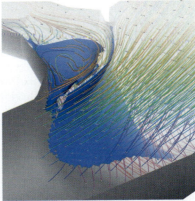

Fig. 7. Flow in Pelton distributor ring. Left: Positive-time FTLE ridges visualizing bifurcation at sickle and a recirculation region. Right: Additionally some positive-time streamlines used for FTLE computation, seeds are visualized by spheres

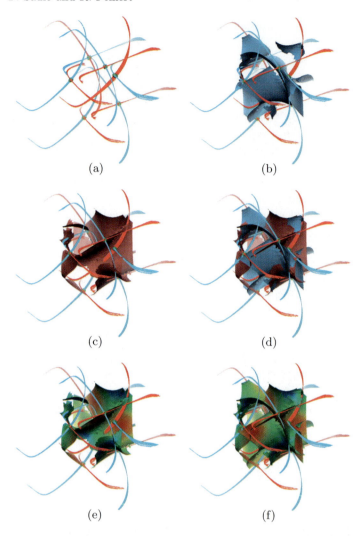

Fig. 8. ABC flow. (a) Seeds around critical points (green spheres), and corresponding streamlines in positive time (red) and negative time (blue). 2D manifolds are 1D-degenerate. (b) Same as (a) with additional positive-time FTLE ridges. (c) Same as (b) but with negative-time FTLE ridges instead of positive-time FTLE ridges (red). (d) Positive-time FTLE ridges (blue) and negative-time FTLE ridges (red). Ridges are well consistent with manifolds. (e) Same as (a) with negative-time FTLE ridges colored with short negative-time FTLE. (f) Additionally positive-time FTLE ridges colored with short positive-time FTLE

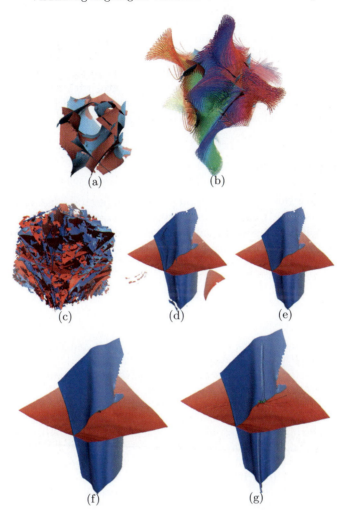

Fig. 9. (a)–(b): another view to the FTLE ridges from Figure 8(d) of the ABC flow. (c)–(g): 3D saddle in Francis draft tube. Positive-time FTLE ridges (blue) and negative-time FTLE ridges (red). (a) FTLE ridges. (b) Additionally positive-time trajectories (arbitrary colors) started from nodes inside the first period of the ABC flow, as used for FTLE computation. Trajectories are well consistent with FTLE ridges. (c) No filtering. (d) Minimum FTLE 3.5 (positive-time) and 4.0 (negative-time). (e) Additionally to (e) suppressing components smaller than 1000 (positive-time) and 4000 triangles (negative-time). (f) Critical point (black) is close to the intersection curve of the two ridges. (g) Seeds around critical point (green) and streamlines in positive time (red) and negative time (blue) from seeds. Streamlines visualizing the 1D manifold of the saddle lie inside positive-time FTLE ridge

FTLE ridges computed at the sickle of the distributor ring. A sickle is a construct where part of the main flow is bifurcated into the injector that forms one of the jets that drive the turbine. One FTLE ridge shows clearly how the flow is split and an other FTLE ridge visualizes a recirculation zone. The FTLE was computed on a $100 \times 100 \times 40$ grid with advection time 0.1 seconds which took 1381 seconds. Ridge regions with FTLE smaller than 22 were suppressed as well as connected components smaller than 2000 triangles.

4 Conclusion

2D height ridges were extracted from 3D FTLE. Several ridge-filtering techniques were proposed in order to suppress noise but also for achieving physically significant visualizations. The ridges were compared to the results from vector field topology, usually resulting in a gain of information and interpretability.

References

1. Asimov, D.: Notes on the Topology of Vector Fields and Flows. Tech. Report RNR-93-003, NASA Ames Research Center (1993)
2. David Eberly: Ridges in Image and Data Analysis. Computational Imaging and Vision. Kluwer Academic Publishers (1996)
3. Furst, J.D., Pizer, S.M.: Marching Ridges. 2001 IASTED International Conference on Signal and Image Processing (2001)
4. George Haller: Distinguished material surfaces and coherent structures in three-dimensional fluid flows. Physica D 149 248–277 (2001)
5. George Haller: Lagrangian coherent structures from approximate velocity data. Phys. Fluids A14 1851–1861 (2002)
6. George Haller: An objective definition of a vortex. J. Fluid Mech. 525 1–26 (2005)
7. James Helman, Lambertus Hesselink: Representation and Display of Vector Field Topology in Fluid Flow Data Sets. IEEE Computer 22(8) 27–36 (1989)
8. Michel Henon: Sur la topologie des lignes de courant dans un cas particulier. Comptes Rendus Acad. Sci. Paris 262 312–314 (1966)
9. De-Bin Huang, Xiao-Hua Zhao, Hui-Hui Dai: Invariant tori and chaotic streamlines in the ABC flow. Phys. Lett. A 237(3) 136–140 (1998)
10. Gordon Kindlmann, Xavier Tricoche, Carl-Fredrik Westin: Anisotropy Creases Delineate White Matter Structure in Diffusion Tensor MRI. In: MICCAI'06, Lecture Notes in Computer Science 3749 (2006)
11. Yuval Levy, David Degani, Arnan Seginer: Graphical Visualization of Vortical Flows by Means of Helicity. AIAA 28(8) 1347–1352 (1990)
12. Lindeberg, T.: Edge detection and ridge detection with automatic scale selection. International Journal of Computer Vision 30(2) 77–116 (1996)
13. Peter Majer: A Statistical Approach to Feature Detection and Scale Selection in Images. Ph.D. thesis, University of Goettingen (2000)

14. Kuangyu Shi, Holger Theisel, Tino Weinkauf, Helwig Hauser, Hans-Christian Hege, Hans-Peter Seidel: Path Line Oriented Topology for Periodic 2D Time-Dependent Vector Fields. Proc. Eurographics/IEEE VGTC Symposium on Visualization (EuroVis '06) 139–146 (2006)
15. Holger Theisel, Tino Weinkauf, Hans-Christian Hege, Hans-Peter Seidel: Saddle Connectors - An Approach to Visualizing the Topological Skeleton of Complex 3D Vector Fields. Proc. IEEE Visualization 225–232 (2003)
16. Holger Theisel, Tino Weinkauf, Hans-Christian Hege, Hans-Peter Seidel: Stream Line and Path Line Oriented Topology for 2D Time-Dependent Vector Fields. Proc. IEEE Visualization 321–328 (2004)

Extraction of Separation Manifolds using Topological Structures in Flow Cross Sections

Alexander Wiebel[1], Xavier Tricoche[2], and Gerik Scheuermann[1]

[1] Institut für Informatik, Universität Leipzig
{wiebel|scheuermann}@informatik.uni-leipzig.de
[2] Scientific Computing and Imaging Institute, University of Utah
tricoche@sci.utah.edu

Summary. The study of flow separation from walls or solid objects is still an active research area in the fluid dynamics and flow visualization communities and many open questions remain. This paper aims at introducing a new method for the extraction of separation manifolds originating from separation lines. We address the problem from the flow visualization side by investigating features in flow cross sections around separation lines. We use the topological signature of the separation in these sections, in particular the presence of saddle points and their separatrices, as a guide to initiate the construction of the separation manifolds. Having this first part we use well known stream surface construction methods to propagate the surface further into the flow. Additionally, we discuss some lessons learned in the course of our experimentation with well known and new ideas for the extraction of separation lines.

1 Introduction

In the panel *"Even more theory, or more practical applications to particular problems: In which direction will Topology-Based Flow Visualization go?"* of the previous workshop (TopoInVis 2005) the (still unsolved) question of benchmark problems arose. These problems should be used to measure the success of the visualization community and more specifically the success of topological methods. The problem posed by one of the participants was the following:

> *"Given a set of data, say of an ICE train, develop a visualization software which is capable to produce these type of Schlichting[1] flow visualizations including the stream surfaces separating from the body,*

---

[1] The speaker refers to drawings in a major textbook for fluid dynamics by Schlichting [9]. See Fig. 1 for an example.

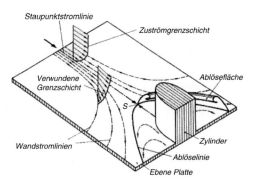

Fig. 1. Drawing of separation line (denoted as Ablöselinie) and separation manifold (denoted as Ablösefläche). Image reprinted from Schlichting [9]

> *including the separation points, vortex core lines and the things which you find in the textbook. [...] it would be really interesting if you could do it in an automatic manner.".*

The fact that this problem was posed as a benchmark problem by a fluid dynamics engineer shows that, despite much research and publications on this topic (see Sec. 2), automatic vortex and separation feature visualization are still problems that lack a satisfactory solution. The separation part of this problem provides the motivation for this paper: we aim at automatically extracting separating stream surfaces, one of the building blocks of the type of representations we just mentioned.

The first and simplest idea for the extraction of separation surfaces is to use previously computed separation lines, i.e. connected locations of flow separation, as seeding curves to integrate stream surfaces in the three-dimensional flow. As we will describe later in the paper, this idea and even some more advanced techniques have limitations. This led us to the development of the method presented in this paper. Namely, we compute the flow projection on cross sections along a separation line, construct the topological skeleton of the resulting planar vector fields, and use the saddle point which appears at the separation locus as a guide for the seeding of the separation manifold near the boundary. Having this first part of the surface we employ a standard stream surface integration scheme to compute the remaining part of the surface.

The contents of this paper are organized as follows. We discuss previous work in section 2. The computation of the topological signature of the separated flow is described in section 3. We then provide results for a variety of realistic CFD problems in section 4. Finally, we conclude our presentation by pointing out topics of future research in section 5.

2 Related Work

As mentioned in the introduction, flow separation has been a topic of central interest for both the theoretical and experimental sides of fluid dynamics research since the 1950's. A complete overview of the topic is clearly beyond the scope of this paper. Therefore we refer the interested reader to [2, 10] and the references therein. Délery [2] described a variety of three-dimensional topological configurations corresponding to separated flows in the smooth setting[2] and associates them with similar experimental visualizations obtained in a wind tunnel. In a very recent paper Surana et al. [10] proposed a formal theory connecting the Navier-Stokes equations to a topological characterization of separation lines and associated 2-manifolds. The same authors also applied their approach to numerical models computed over simple geometries in [11]. Unfortunately the topological characterization advocated by these authors fails to extract the separation lines and manifolds present in CFD flows defined over more complex polygonal geometries if only the spatial velocity and vorticity data is provided and no access to the simulation is possible (precomputed data). In this case it may be impossible to locate the saddles needed for three of the four separation types mentioned by Surana et al. [10]. This is especially true for the delta wing example they mention. A higher order saddle (or several very close saddles) has to exist at the tip of the wing to serve as origin for separation lines. It strongly depends on the discretization (at the pointed tip) whether such a saddle can be found. Additionally, the skin friction field derived from the precomputed data may be of minor quality prohibiting the use of the formulas reported by Surana et al. [10].

From the visualization viewpoint, the extraction of line-type flow features has motivated a significant body of research in recent years. In addition to the definition and computation of vortex core lines (see [8] for a bibliography), several authors have attempted to detect and display separation and attachment lines on surfaces. The major contribution in this field was made by Kenwright [6] who proposed a method that looks for the presence of separation lines on a cell-wise basis. Upon the assumption of local linearity of the flow a section of separation line is extracted in each triangle as the intersection of specific lines present in the phase portrait of a first-order critical point with the interior of the cell. The discontinuity across cell edges yields disconnected line segments, which was addressed in a subsequent paper [7]. Making the observation that the criterion used in the original method is in fact equivalent to that of zero streamline curvature the authors reformulated their extraction algorithm as the computation of isolines of the flow curvature field. The resulting curves must be filtered in a post-processing step to discard false positives (see [8]). Moreover, the requirement of zero curvature is far too restrictive to account for the general flow geometry associated with flow

---

[2] Here "smooth setting" means the opposite of "discrete setting" like for vector fields from CFD simulations.

separation or attachment. In fact, this criterion was basically tailored to the special case of delta wings, for which separation and attachment lines describe almost straight lines linking the tip of the wing to its back edge. Based on this observation Tricoche et al. [13] recently proposed a method combining a local predictor (the flow divergence) and a global corrector (the one-dimensional convergence behavior of streamlines) to provide a more flexible and robust extraction mechanism. Yet, the corresponding method requires the integration of many streamlines to ensure reliable detection of flow convergence and the filtering of false positives remains a non-trivial and error-prone task. Concerning the use of cross-flow sections or cutting planes Tricoche et al. [12] and Wu et al. [14] proposed methods related to ours. Tricoche uses the cutting planes for general flow visualization problems on special application data sets while Wu discusses cross sections explicitly in the field of flow separation from the fluid dynamics side.

3 Extracting Separation Manifolds

In this section we propose an approach for seeding stream surfaces representing sheets that divide the flow at separation and attachment lines, i.e. separation manifolds. First, we discuss the problems usually arising when trying to compute such surfaces then we give possible solutions to the problems and finally we describe the basic procedures involved in our solution.

3.1 Seeding Stream Surfaces

Stream surfaces are well known from the literature [4] and good implementations [3] are readily available for visualization of vector fields. The quality of a stream surface or even the possibility to compute it, however, strongly depends on the availability of a good seed curve for the integration of the dense set of streamlines spanning the surface. The first and obvious idea to get such line strips is to use the extracted feature lines. Unfortunately, this is only a good idea at first sight, because the integration of streamlines in this case starts directly from the surface. This poses problems resulting from the discretization of the surface with polygons and from numerical inaccuracies. The first problem means that if we start streamlines directly from the surface, the integration may yield steps that lead out of the grid as the surface has discontinuities at polygon edges and is not smooth. The second problem is that vertices of feature lines may lie on the wrong side of the surface polygon, i.e. outside of the grid, because of small numerical round-off added while computing the line segments.

Our first attempt to solve these problems consisted in moving the feature lines by a small distance in the direction of the surface normal in order to avoid the issues caused by the irregularities of the polygonal surface while remaining in its immediate vicinity. When we started stream surfaces from the shifted lines the integration ran well, but the surfaces, in some cases, did not capture the separation manifold correctly. This is due to the fact that the separation need not happen perpendicular to the surface. Therefore, this method requires knowledge of the angle of the separation in relation to the surface normal. Assuming that this information is available for a point on a feature line we would be able to determine the correct offset vector for this point. We found a formula to do this in the previously mentioned paper of Surana et al. [10]. However, we chose an alternative solution that we found simpler and for our discrete numerical data more appropriate:[3] we chose to compute the topology of the projected flow in cutting planes along the feature lines and to use the separatrix indicating the separation as guide for the movement of our feature lines or, more precisely, the generation of new starting line strips. Near the surface, the angle between the separation and the surface normal is the same as the angle of this separatrix. The following subsection will describe this idea in more detail.

3.2 Moving Cutting Planes

Cutting planes, a very basic and widely used technique, gain more informational value if they are moved along interesting curves (see Fig. 4a). In our case, the type of curves to be used is inherent in the idea of using moving cutting planes. The separation lines provide the natural paths for the moving planes but also leave several degrees of freedom for the orientation of the plane.

Plane Orientation

The orientation of the plane is a critical parameter of our flow exploration technique because the topological structure in the plane can change dramatically when changing the orientation of the plane. Even when keeping the center position of the plane constant, features can appear or disappear for different orientations of the plane. We would like to use the topology in the plane as a guide for the structure of the flow separation. As the structures which are important for us appear normal to the flow along the separation, we use the surface flow vector at the separation line as first approximation of the normal vector for the cutting plane (see Fig. 4). With this choice, the projected vector field should exhibit a zero vector exactly where the separation line intersects the plane. The zero is a half-saddle with one separatrix leading away from the surface. Near the surface this separatrix indicates the direction of the separation. However, the saddle point can only be found reliably in an analytical setting. As our procedure requires resampling the data and projecting it into the plane, we introduce numerical noise and round-off error to some extent. This can cause the saddle to move. Lying directly on the boundary

---

[3] As mentioned before, we are only working on precomputed data. Thus, not all quantities used in the formula by Surana et al. [10] are available.

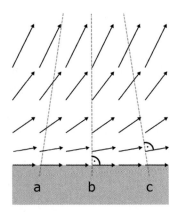

Fig. 2. Saddle moves when tilting plane in separating flow: **a)** no saddle is present because flow is nowhere perpendicular to plane, **b)** half-saddle because plane is perpendicular to wall and thus to surface flow, **c)** saddle moved a small distance away from the wall because plane is perpendicular to separating flow

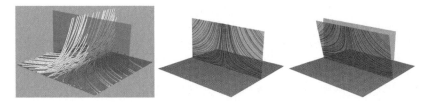

Fig. 3. The left image shows streamlines in a synthetic vector field representing perfect separation and an example for a cutting plane. The image in the middle gives an impression of the projected flow in the plane and the right image shows the flow after tilting the plane. In the middle image only the upper part of the saddle can be seen. The saddle lies exactly on the base plane bounding the flow from below. All sectors of the saddle are observable after it is shifted in the lower image

of the body, even a small shift can cause the saddle to move out of the plane and thus disappear. Hence, in the case where we do not find a saddle point in the plane we have to adjust its orientation in order to move the desired saddle back onto the plane. Figure 2 shows the basic idea of the adjustment and Figure 3 gives an impression for a real 3D vector field. While we have to tilt the plane against the flow direction for separating flow, we must tilt it with the flow to move the saddle into the plane for attaching flow (see Fig. 2).

We said previously that the orientation of the plane can change the flow dramatically. We have to keep this in mind when tilting the plane, i.e. we have to change the orientation of the plane as little as possible. Changing the inclination of all cutting planes along a line in the same way is the simplest idea but it does not account for the different flow situations at the different base positions along the line. Additionally, such a uniform change would surely not be the smallest change possible for all planes at the same time. Hence,

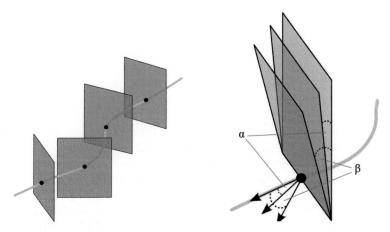

Fig. 4. a) Cutting Planes moving along feature line oriented *normal to the flow* at their center position. **b)** Tilting the plane by tilting the normal vector in the plane spanned by the normal vector of the plane and the normal vector of the surface

we do not change the inclination uniformly but instead compute the angle for each separate plane through an iterative approach.

As mentioned earlier, we use the velocity as first approximation for the plane normal. To tilt the plane, we turn its normal vector in the plane spanned by the wall normal and the plane normal. This is illustrated in Figure 4b. The iterative approaching works in two parts. All steps are described in the following:

1. a) Set current angle to a small and constant user prescribed angle.
   b) Tilt the original plane by the current angle (against the flow for separation and with the flow in the attachment case).
   c) Compute the flow in the tilted plane.
   d) Extract the topological structures of the flow.
   e) If there is no saddle, decrease the current angle and go back to 1b.
   f) If there is a saddle go to 2a.
2. a) Set α_{max} to the current angle and $\alpha_{min} := 0$.
   b) Decrease the current angle.
   c) Tilt the original plane by the current angle.
   d) Compute the flow in the tilted plane.
   e) Extract the topological structures of the flow.
   f) If there is a saddle set α_{max} to the current angle.
   g) If there is no saddle set α_{min} to the current angle.
   h) Set the current angle value to $0.5(\alpha_{max} + \alpha_{min})$ and go to 2c.

We iterate for both parts until a maximum number of iterations has been reached. The first part tries to find an initial angle for which a saddle is present on the plane. In the second part of the algorithm, we know that there is a saddle in the plane and try to find the plane with smallest tilting angle *and* a saddle.

3.3 Construction of Seed Line for Separation Manifold

In the previous subsection we described how we compute the topological signature of the separation in the cutting planes. Having this signature, i.e. the separatrices of the saddle point, we can use it to construct the first part of the separation manifold and a start strip for the rest of the surface. At first we have to choose the correct separatrix out of the four separatrices emanating from the saddle point. For this task, we take the vector representing the first segment of each of the separatrices and compute its dot product with the body surface normal (all vectors normalized). The segment with the largest value indicates the desired separatrix, as this separatrix is the one which *leaves* the surface.

To construct the first part of the surface, we take a number of steps along the chosen separatrices in all planes. The step size and the number of steps m can be determined by the user. We then connect the steps of neighboring planes to construct triangles. These triangles build up a surface consisting of m rows of triangles. This surface is the first part of the separation manifold. Splitting the distance for moving along the separatrix into several steps is only important for constructing the triangles. The upper boundary of the last row of triangles, i.e. the connection of the (last) step on all separatrices, constitutes the starting line strip for the rest of the separation manifold.

As mentioned before when discussing the tilting of the plane, it is important to use only structures of the flow in the planes that are very close to the separation line and thus to the body surface. This ensures that the separatrices precisely approximate the structure of the separation manifold.

4 Results

In this section we demonstrate our method on two different datasets from CFD simulations. We give a short overview of each dataset, apply our method and discuss the resulting images.

4.1 Blunt Fin

Our first example is a standard reference for flow visualization, the well known blunt fin dataset [5], courtesy of NASA. It represents a steady Mach 2.95 airflow over a flat plate with a blunt fin rising from the plate. The free inflow with a Reynolds number of $2.1 \cdot 10^6$ is parallel to the plate and to the flat part of the fin. The dataset represents only one half of the real flow, as it is assumed to be symmetrical about center plane of the fin. In front of the fin two horseshoe vortices coexist with a shock front. The first vortex causes the flow reaching it to separate from the plate. This separation is what we are interested in. The upper image in Figure 5 shows a LIC [1] visualization of

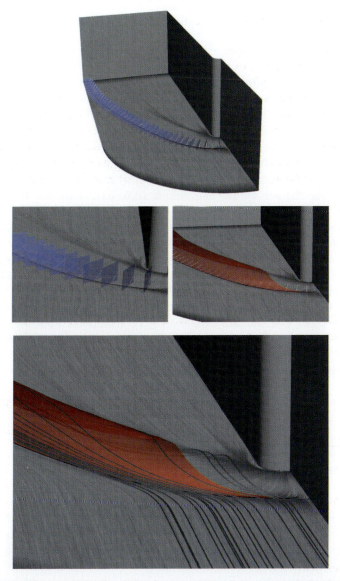

Fig. 5. Blunt fin dataset: The upper image gives an overview of the dataset, the separation line and a number of cutting planes with their topological skeletons. A close-up of the same setting is provided in the middle left image. The middle right and the lower images show the separation manifold constructed from the saddles. In the middle right image the surface is compared with the separatrices in the cutting planes. Streamlines in the lower image prove that the extracted surface is indeed the separation manifold

Table 1. Final angle of plane and distance of saddle to surface. Note: size of data set is approx. 10 units

	minimum	maximum	average
Angle (°)	0.006	0.27	0.026
Distance	0	0.008	0.0016

the flow over the plate where patterns of converging flow indicate the separation. For the application of our method we extracted a separation line (shown in black) and computed the topological skeleton of 38 section planes moving along the separation line, which took us 2 minutes. We show the section planes to illustrate the procedure. The closeup in the middle left image shows how the separatrices of the saddles in the planes indicate the separation.

The middle right and the lower images show a red surface representing the separation manifold we constructed from the cutting planes. In the right image it is combined with the separatrices of the saddles in the cutting planes. The separatrices are tangential to the surface near the saddle point but intersect it when the distance to the saddle increases. This supports what we mentioned before: the separatrices are good guides for the surface construction but only very close to the saddle. Finally, streamlines in the lower image prove that we computed the correct stream surface, i.e. the one representing the separation manifold.

4.2 ICE High Speed Train

The second dataset is a more practically relevant example. It is the high-speed (*ICE*) train already mentioned in the introduction. This dataset is the result of a simulation of the train traveling at a velocity of about 250 km/h with wind blowing from the side at an angle of 30 degrees. The wind causes vortices to form on the lee side of the train, creating a drop in pressure that has adverse effects on the train's track holding. The vortices and the flow passing over the top of the train lead to separating flow at the upper angle on the lee side. The separation line we extracted and the flow structure in the corresponding cutting planes is shown in Figure 6. Note how there are no saddles along the blue line where the line does not capture the location of separation correctly. We observed this also in cases where the separation becomes very weak or fuzzy along a line. Streamlines, the separation line, the separation surface we seeded using the cutting planes, and a LIC on the surface of the train give a picture of the complete situation in the lower image. The causal connection between the separation and the vortex formation becomes obvious in this image.

The computations for the 100 cutting planes took 12 minutes.

Fig. 6. The upper images show the surface of the ICE train with one separation line and the topological skeletons of about 100 cutting planes along the separation line. Blue and green points mark sinks and source in the cutting planes. The lower image shows a separation manifold generated from the guides in the cutting planes. We additionally provide streamlines to prove that we found the correct separation manifold

Table 2. Final angle of plane and distance of saddle to surface. Note: length of train is approx. $5 \cdot 10^4$ mm

	minimum	maximum	average
Angle (°)	$< 10^{-8}$	3.26	0.60
Distance (mm)	0	82.04	16.75

5 Conclusion

We have presented a method for automatic computation of separation manifolds from separation lines on bodies immersed in a flow. The method constructs and uses flow cross-sections and the topological signature of the separation therein to construct the surface section in the direct vicinity of the boundary. The construction of the remaining part of the surface relies on standard techniques. Our method proved its usefulness and robustness through application to several different CFD datasets.

For datasets with more intricate flow structures than those presented in this paper, e.g. flows with a very small separation angle, we found problems with the cross-flow in planes very close to the boundary. Thus, our future work will include close collaboration with engineers in order to inspect what is happening near the boundary in these simulations, to try to understand what the problems are and how we can improve our method to handle these problems. As mentioned previously the quality of the surfaces representing the separating flow strongly depends on the accuracy of previously extracted separation lines. Research in the direction of separation line extraction is thus still necessary and will be part of our work until extraction can be performed efficiently and reliably.

Acknowledgments

The authors wish to thank Markus Rütten from DLR Göttingen for providing the ICE dataset. Further we are thankful to all members of the FAnToM development team for their programming efforts. This work was partly supported by DFG grant SCHE 663/3-7.

References

1. Brian Cabral and Leith Casey Leedom. Imaging Vector Fields Using Line Integral Convolution. In *SIGGRAPH '93: Proceedings of the 20th Annual Conference on Computer Graphics and Interactive Techniques*, pages 263–270, New York, NY, USA, 1993. ACM Press.
2. Jean M. Délery. Robert Legendre and Henri Werlé: Toward the Elucidation of Three-Dimensional Separation. *Ann. Rev. Fluid Mechanics*, 33:129–154, 2001.
3. Christoph Garth, Xavier Tricoche, Tobias Salzbrunn, Tom Bobach, and Gerik Scheuermann. Surface Techniques for Vortex Visualization. In Oliver Deussen, Charles Hansen, Daniel A. Keim, and Dietmar Saupe, editors, *Data Visualization 2004 - Eurographics/IEEE TCVG Symposium on Visualization Proceedings*, pages 155–164, Konstanz, Germany, May 2004. Eurographics Association.
4. Jeffrey P. M. Hultquist. Constructing Stream Surfaces in Steady 3D Vector Fields. In Arie E. Kaufman and Gregory M. Nielson, editors, *Proceedings of the 3rd conference on Visualization '92*, pages 171–178, Boston, MA, 1992.
5. C. M. Hung and P. G. Buning. Simulation of blunt-fin induced shock wave and turbulent boundary layer separation, AIAA Paper 84-0457. In *AIAA Aerospace Sciences Conference*, Reno, NV, January 1984. AIAA.
6. D. N. Kenwright. Automatic detection of open and closed separation and attachment lines. In IEEE Computer Society Press, editor, *IEEE Visualization Proceedings*, pages 151–158, Los Alamitos, CA, 1998.
7. D. N. Kenwright, C. Henze, and C. Levit. Features extraction of separation and attachment lines. *IEEE Transactions on Visualization and Computer Graphics*, 5(2):135–144, 1999.

8. M. Roth. *Automatic Extraction of Vortex Core Lines and Other Line-Type Features for Scientific Visualization*. PhD thesis, ETH Zürich, 2000.
9. Herrmann Schlichting. *Grenzschicht-Theorie*. Braun, Karlsruhe, 1992.
10. A. Surana, O. Grunberg, and G. Haller. Exact Theory of Three-dimensional Flow Separation. Part I: Steady Separation. *J. Fluid Mech.*, 564:57–103, 2006.
11. A. Surana, G. B. Jacobs, and G. Haller. Extraction of Separation and Reattachment Surfaces from 3D Steady Shear Flows. *AIAA Journal, in press*, 2006.
12. Xavier Tricoche, Christoph Garth, Gordon Kindlmann, Eduard Deines, Gerik Scheuermann, Markus Ruetten, and Charles Hansen. Visualization of Intricate Flow Structures for Vortex Breakdown Analysis. In Holly Rushmeier, Greg Turk, and Jarke J. van Wijk, editors, *Proceedings of the IEEE Visualization 2004 (VIS'04)*, pages 187–194. IEEE Computer Society, October 2004.
13. Xavier Tricoche, Christoph Garth, and Gerik Scheuermann. Fast and robust extraction of separation line features. In *Scientific Visualization: The Visual Extraction of Knowledge from Data*, pages 245–263. Mathematics + Visualization, Springer, 2005.
14. J. Z. Wu, R. W. Tramel, F. L. Zhu, and X. Y. Yin. A vorticity dynamics theory of three-dimensional flow separation. *Physics of Fluids*, 12:1932–1954, August 2000.

Topology Based Selection and Curation of Level Sets

Chandrajit Bajaj, Andrew Gillette, and Samrat Goswami

Center for Computational Visualization, University of Texas at Austin Austin, Texas 78712
bajaj@cs.utexas.edu, agillette@math.utexas.edu, samrat@ices.utexas.edu

Summary. The selection of appropriate level sets for the quantitative visualization of three dimensional imaging or simulation data is a problem that is both fundamental and essential. The selected level set needs to satisfy several topological and geometric constraints to be useful for subsequent quantitative processing and visualization. For an initial selection of an isosurface, guided by contour tree data structures, we detect the topological features by computing stable and unstable manifolds of the critical points of the distance function induced by the isosurface. We further enhance the description of these features by associating geometric attributes with them. We then rank the attributed features and provide a handle to them for curation of the topological anomalies.

1 Introduction

Problem and Motivation

The selection of isosurfaces (or alternatively level sets of a trivariate function) for the visualization of volumetric imaging or simulation data, is often subject to the constraints imposed by the application domain. While some applications put more stringent criteria on this process than others, every application requires an analysis of topological and geometrical information to ensure that an appropriate choice is made. Topological constraints require that the extracted level set should have a certain topology. Naturally, this problem has drawn the attention of researchers for a long time. As a result, the basic topological information of level sets can be unambiguously encoded by the well-known data structure, called the Contour Tree [31]. The Contour Tree (CT) is a topological description of the entire volumetric data, and furthermore encodes the information of the number of connected components.

While extremely useful, CT does not capture all the necessary information about the topology of the level sets. In particular, information about the combinatorics and topology of tunnels (complementary space) of every connected component of the level set is not encoded in CT. To overcome this deficiency, an enhanced CT data structure was proposed by Pascucci and Cole-McLaughlin [29], who gave an algorithm to further detect and store some additional information of the level set, in an Augmented Contour Tree (ACT).

The ACT, while appropriate in encoding the topology related to the first 3 Betti numbers [29] of each level set, it does not give a quantitative measure of the level set and its complementary space features, to aid in its selection or curation . This is the main motivation of our work. In this paper, we will show that suitable selection of an isosurface in any application context must be additionally guided by a close examination of the complementary space to the primal domain. Further, we will show how relevant topological and geometrical features of a selected isosurface can be calculated, ordered, visualized, and, if necessary, curated.

The significance of such complementary space features can be seen in the isosurface choices for the simple example of the ion channel Gramicidin A, obtained from soil bacteria Bacillus brevis. The initial molecular surface selection process makes use of the CT. In Figure 1 (a.1-4) we show two isosurfaces extracted from the same edge of the contour tree for the volumetric data which has been synthetically created from the atomic model obtained from Protein Data Bank (PDB) [5]. The PDB entry of this molecule is 1MAG. Figure 1 (a.1) shows a snapshot of the volume rendering of the data in our in-house interrogative volume visualization software VOLROVER [13] along with the CT shown in the bottom panel. In Figure 1 (a.2) we show the molecular surface selected using CT which does not accommodate any tunnel through the molecular surface. Figures 1 (a.3,4) show two views of another molecular surface selected from the same scalar volume which, on the other hand, has the ion channel through the molecule present. This means the choice of the first isosurface is incorrect. Note that there is no change in the number of components as one contour is deformed into the other, hence the CT is not sufficient in guiding our choice of isovalue. Further as can be seen in Figure 1 (a.4), the surface accommodates two more small tunnels (marked by black circles) which are merely the artifacts of the selection process. Using the algorithm presented in this paper, we can eliminate those small tunnels.

Figure 1 (b.1-3) shows another example of the isosurface of the molecule mouse Acetylcholinesterase (mAchE). In this case, the active site of the molecule is buried deep inside a depression on the molecular surface. We will later computationally define such depressions and we will call them "pockets". Correct extraction of the molecular surface of mAchE, in this case, requires that the pocket near the active site is properly preserved and also it should not be too narrow, e.g. narrower than the diameter of Acetylcholine molecule which has to pass through the opening of the pocket in order to bind to the active site of mAchE. Figure 1 (b.1) shows all such pockets identified on the molecular surface of mAchE. The small pockets on the surface arising from incorrect selection of isovalue, are not desired because they adversely affect the calculation of molecular energetics. Using the algorithm presented in this paper, we can rank the pockets based on their geometric attributes (volume, for example) and that helps deleting those small pockets. Figure 1 (b.2) shows the surface after selective removal of all the small pockets keeping only three significant ones which include the one near the active site. Closeup of the main pocket is shown in Figure 1 (b.3).

The selection of an isosurface should also aim to preserve inherent symmetry in the data. In Figure 1 (c.1-3) we show two isosurfaces for the Nodavirus. This icosahedral virus infects the central nervous system of fish and causes a disease called viral nervous necrosis. The main point of contention while selecting an isosurface

Topology Based Selection and Curation of Level Sets 47

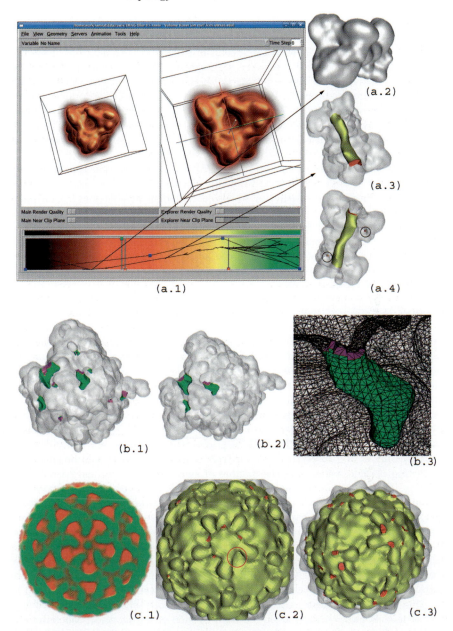

Fig. 1. Three examples of molecular surface selection are shown. (a.1-4) shows the selection of Gramicidin A that preserves the ion channel. (b.1-3) shows the selection of molecular surface for mouse Acetylcholinesterase (mAchE) where the pocket near the active site for binding Acetylcholine is preserved. (c.1-3) shows the selection of isosurface of the viral capsid of Nodavirus that brings out the inherent symmetry. In all three examples, we show how our algorithm for detecting and evaluating topological features aids the proper isosurface selection

is preserving the symmetry of the viral capsid. We show the volume rendering of the cryo-EM density map of the virus in Figure 1 (c.1). Figure 1 (c.2) shows one plausible isosurface that does not respect the symmetry. In particular, near the points of five-fold rotational symmetry, we ought to have five tunnels arranged in a circle or none at all. In Figure 1 (c.3) we show another isosurface which actually reflects the symmetry properly. Without careful investigation of the complementary space topology, aided by the algorithm described in this paper, it is not possible to apply the knowledge of these features to guide the isosurface selection. Therefore, we seek both topological and geometrical details of the complementary space to an extracted isosurface.

The overall isosurface extraction process from a scalar volume is thus guided by the knowledge of both topology and domain-specific geometry of both the primal and complementary space. As in [19], we also consider the homology as a measure of topological complexity of the extracted level set. We further consider the depressions on the surface as added complexity as they play a bio-chemically vital role in the context of selection of molecular interface [18]. We call these depressions *pockets*. The main contribution of this work is the attachment of geometry and possibly other domain-specific attributes with the detected topological features that guides the subsequent curation process.

Prior work

Systematic interrogation of topological and geometric attributes of the level sets over a range of isovalues started with the introduction of two powerful data structures Contour Spectrum (CS) [4] and Contour Tree (CT) [31]. While CS focuses mostly on the differential and integral attributes like, area, volume, curvature etc., CT encodes topological properties like, number of connected components etc., of the contours over a range of isovalues. Historically, CT was first introduced in [6]. It was first used with regard to isosurfaces by van Kreveld et al. [31]. They also used CT to compute seed sets, which help generate isosurfaces efficiently. Further work by Carr et al. proved that contour trees of any dimension could be computed in $O(n \log n)$ time, where n is the number of simplices in the geometrical decomposition [8]. Pascucci and Cole-McLaughlin expanded the topological information in the CT data structure by adding Betti number information to each edge of the graph [29]. Recently, CT has been further enhanced by adding information about domain specific geometric attributes and such multi-attributed CT (MACT) proved to be very useful in analyzing the bio-chemical properties of the molecular interfaces [35]. Carr, Snoeyink and van de Panne simplified the CT based on local geometric data [9]. The algorithm in [9], although different from our approach, can be used to further facilitate in simplification of the volume and selection of the isosurface.

Due to the fundamental nature of the problem, topology simplification of geometric models have received attention from outside of visualization community as well. In order to achieve controlled topological simplification of triangulated geometry, Guskov and Wood [25] grow a small ball of radius ϵ on the surface to detect small tunnels and remove them by cutting the mesh. Wood et al. [32] use the Reeb graph of a *height* function to detect and delete small handles from an isosurface. El-Sana and Varshney have worked on topology controlled simplification of CAD models where they first detect the crease edges and roll a ball of small radius to identify the holes which do not allow the ball to pass through [20]. Nooruddin and

Turk have proposed an algorithm that converts a model into volumetric data and apply dilation and erosion to perform simplification [28]. It is to be noted, that all these techniques fall short of depicting the symmetry and a proper ranking of the geometry of the depressions and tunnels of the geometry to be simplified.

A fundamentally different approach was due to Edelsbrunner et al. [19] who proposed the notion of *persistence* in the context of the alpha complex to detect topological features which prevail while the alpha complex undergoes a filtration. A series of results followed along the same line which formalized the notion of persistent homology in order to distinguish between stable topological features from unstable or noisy features [7, 12, 37, 36]. A similar notion of persistence has also proved to be useful in detecting short-lasting and noisy topological features in the context of witness complex [15]. At this stage it is important to note the novelty of our approach. We compute the topological features related to the homology group of the level set and attach geometric attributes which are often meaningful in the context of application the scalar volume has originated from.

The key ingredient of our algorithm in ranking the topological features of the extracted level set is the *distance function* over $\mathbb{R}^3$. The distance function has been used earlier for reconstruction and image feature identification [1, 10, 17, 21]. Chazal and Lieutier [11] have used it for stable medial axis construction. Dey, Giesen and Goswami have used distance function for object segmentation and matching [16]. Goswami, Dey and Bajaj have used it for detailed feature analysis of shape via an annotation of flat and tubular features in addition to shape segmentation [22]. Recently, Bajaj and Goswami have shown a novel use of distance function, induced by a molecular surface, in order to detect secondary structural motifs of a protein molecule [2]. The close connection between the critical point structure of the distance function and the topology of the surface, and its complement, is what we utilize to detect and remove small topological artifacts.

Approach

The main contribution of this work is the systematic use of the distance function induced by an isosurface, to geometrically complement the encoding of the topology by the Contour Tree, in yielding a curated, selection. With our new approach the selected isosurface is extracted, and then filtered, with the aid of the critical point structure of the distance function, which allows detection and a geometrical ranking of the complementary structure of the isosurface, i.e. the *tunnels* and *pockets*.

First, a suitable isovalue is selected using CT in order to select an isosurface with the required number of components. In the case of molecular interface selection, the number of components is always one. The subsequent computations based on the distance function are then applied to detect the tunnels and pockets. Finally, these features are ranked according to some domain-aware "importance" function which usually quantifies the geometric attributes of those features, and thereby allows the removal of insignificant ones.

We first give a brief description of the distance function, here. Given a compact surface Σ smoothly embedded in $\mathbb{R}^3$, a distance function h_Σ can be designed over $\mathbb{R}^3$ that assigns to each point its distance to Σ.

$$h_\Sigma : \mathbb{R}^3 \to \mathbb{R}, \quad x \mapsto \inf_{p \in \Sigma} \|x - p\|$$

In this context, Σ is the level set. For the ease of computation, we approximate h_Σ by h_P which assigns to every point in $\mathbb{R}^3$, the distance to the nearest point from the set P which finitely samples Σ.

$$h_P\ :\ \mathbb{R}^3 \to \mathbb{R},\ \ x \mapsto \min_{p \in P} \|x - p\|$$

We identify the maxima and index 2 saddle points of h_P which lie outside the level set. The stable manifolds of these critical points help detect the tunnels and the pockets of Σ. Additionally these stable manifolds are used to compute geometric attributes of the detected topological features that they correspond to. Thus we obtain a description of the isosurface, and its complement, in terms of its topological features quantified by their geometric properties, based on which the insignificant features are removed.

2 Preliminary

2.1 Contour Tree

Isosurfaces and contour trees are derived from scalar fields. A scalar field can be characterized as a domain M and a function $f : M \to \mathbb{R}^1$. In differential topology and Morse theory, the critical values of f are formally defined as those values $r \in \mathbb{R}^1$ for which the derivative map df_x is not surjective for some point $x \in f^{-1}(r)$ (see [24], for example) Put differently, r is a critical value of f if and only if $f^{-1}(r)$ is not a manifold of dimension $\dim(M) - 1$. Each level set $f^{-1}(r)$ is a collection of contours and the topology of these contours allows us to create the contour tree. We note that in all of our examples, our domain M will be $\mathbb{R}^3$, ensuring that our data structure is in fact a tree and not the more general Reeb graph.

The unqualified term "contour tree" refers to a data structure created from a subset of the critical values of f. The first such data structure to be computed efficiently was a "minimal" contour tree by de Berg and van Kreveld [14]. A minimal contour tree has nodes only for isovalues at which contours emerge, split, merge, or vanish. The edges of a minimal contour tree connect nodes along which a contour smoothly deforms and hence indicate the evolution of a contour over a range of isovalues. A minimum contour tree of any dimension can be computed in $O(n \log n)$ time as was proved by Carr et al. in [8]. Such trees can be used to compute seed sets, that is, a set of points from which all contours of a particular level set can be generated [31].

To capture more topological information, the augmented contour tree, as defined by Carr, Snoeyink and Axen, was introduced in [8]. In their terminology, the augmented contour tree refers to a contour tree with nodes for all values in the range of the scalar field, not just the critical values. Thus, the augmented or "full" contour tree can be reduced to the minimal contour tree by removing all degree two nodes. Pascucci and Cole-McLaughlin expanded the data structure by attaching the Betti numbers of each contour in a level set to its corresponding edge in the "full" contour tree [29]. The Betti numbers of a surface, however, are a strictly topological feature and thus do not indicate the geometrical significance of the tunnels and voids that they count. Moreover, it is not clear how to use this data structure in order to selectively remove some undesired topological artifacts.

2.2 Voronoi-Delaunay

In this paper we always assume the distance metric to be Euclidean unless otherwise stated. For a finite set of points P in $\mathbb{R}^3$, the Voronoi cell of $p \in P$ is

$$V_p = \{x \in \mathbb{R}^3 \,:\, \forall q \in P - \{p\},\, \|x - p\| \leq \|x - q\|)\}.$$

If the points are in general position, two Voronoi cells with non-empty intersection meet along a planar, convex Voronoi facet, three Voronoi cells with non-empty intersection meet along a common Voronoi edge and four Voronoi cells with non-empty intersection meet at a Voronoi vertex. A cell decomposition consisting of the *Voronoi objects*, that is, Voronoi cells, facets, edges and vertices is the Voronoi diagram Vor P of the point set P.

The dual of Vor P is the Delaunay diagram Del P of P which is a simplicial complex when the points are in general position. The tetrahedra are dual to the Voronoi vertices, the triangles are dual to the Voronoi edges, the edges are dual to the Voronoi facets and the vertices (sample points from P) are dual to the Voronoi cells. We also refer to the Delaunay simplices as *Delaunay objects*.

2.3 Critical Points of h_P

The distance function h_P induces a flow at every point $x \in \mathbb{R}^3$. This flow has been characterized earlier [21, 22]. See also [17]. For completeness we briefly mention it here.

The critical points of h_P are the points in $\mathbb{R}^3$ which lie within the convex hull of its closest points from P. It was shown by Siersma [30] that the critical points of h_P are the intersection points of the Voronoi objects with their dual Delaunay objects (Figure 2).

- *Maxima* are the Voronoi vertices contained in their dual tetrahedra,
- *Index 2 saddles* lie at the intersection of Voronoi edges with their dual Delaunay triangles,
- *Index 1 saddles* lie at the intersection of Voronoi facets with their dual Delaunay edges, and
- *Minima* are the sample points themselves as they are always contained in their Voronoi cells.

In this discrete setting, the index of a critical point is the dimension of the lowest dimensional Delaunay simplex that contains the critical point.

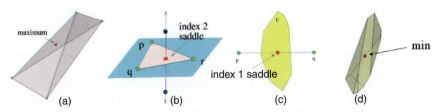

Fig. 2. The relative position of Voronoi and their dual Delaunay objects that results in the generation of critical points

At every $x \in \mathbb{R}^3$, a unit vector can be assigned that is oriented in the direction of the steepest ascent of h_P. The critical points are assigned zero vectors. This vector field, which may not be continuous, nevertheless induces a flow in $\mathbb{R}^3$. This flow tells how a point x moves in $\mathbb{R}^3$ along the steepest ascent of h_P and the corresponding path is called the *orbit* of x.

For a critical point c its stable manifold is the set of points whose orbits end at c. The stable manifold of a maximum is a three dimensional polytope whose boundary is composed of the stable manifolds of the index 2 saddle points which in turn are bounded by the stable manifolds of index 1 saddle points and minima. See [16, 21] for the detailed discussion on the structure and computation of the stable manifolds of the critical points of h_P.

2.4 Betti Numbers

The i-th Betti number of a manifold is formally defined as the rank of its i-th homology group, H_i. Homology groups are quotient groups; H_i is the i-th cycle group modulo the i-th boundary group. Therefore, H_i is the free abelian group generated by cycles of i-chains that are not boundaries of $(i+1)$-chains. Hence, the i-th Betti number counts the number of independent (i.e. non-homologous) non-bounding cycles. Based on these definitions, we have the following informal notions of Betti numbers for 2-manifolds. The 0-th Betti number equals the number of connected components, the 1-st Betti number equals twice the number of through holes, and the 2-nd Betti number equals the number of voids. For an isosurface (or in general a 2-manifold) only the first three Betti numbers can be non-zero.

3 Algorithm

In this section, we describe an algorithm that detects the tunnels and pockets using the critical point structure of the distance function.

3.1 Sampling of Level Set

In order to successfully apply the critical point structure of the discrete approximation of the distance function h_Σ by h_P, we require a suitable discrete approximation of the level set. Apparently primal contouring (Marching Cubes [27]) and dual contouring [26] are good choices to extract a discrete approximation of the level set from the scalar volume. Although variants of these approaches have been researched extensively to produce a topologically consistent isosurface, the main disadvantage lies in the fact that the sampling of the extracted surface is oblivious to the feature. Note, we need a set of points P to approximate h_Σ by h_P, and we also need h_P to follow h_Σ closely so that we do not miss the topological features of Σ in this process of translating it to the discrete setting. Recently, we have developed an algorithm which ensures that the discretization of the level set has sufficiently dense sampling for it to be a subcomplex of the Delaunay triangulation of the set of samples. This guarantees that the sampling is feature-sensitive and therefore the discretization follows closely the distance function induced by the true level set. Due to space limitation, we omit the details of the algorithm here and refer the reader to [23].

3.2 Classification and Clustering of Critical Points

The critical points of h_P are detected by checking the intersection of the Voronoi and its dual Delaunay diagram of the point set P sampled from Σ. The critical points are primarily of three types depending on if the Voronoi/Delaunay object involved lies interior, exterior to Σ, or if the Voronoi object crosses Σ. The maxima can not lie on the surface and therefore they are only of two types - interior and exterior. The minima are sample points themselves and therefore they are always on Σ. The saddle points can be any of three types mentioned above.

We use C_2 to denote the set of index 2 saddles which is partitioned into three classes $C_{2,I}$, $C_{2,O}$ and $C_{2,S}$. The set of maxima is denoted as C_3 which is partitioned into two classes $C_{3,I}$ and $C_{3,O}$. Using the hierarchical nature, we build an *incidence graph* over $C_2 \cup C_3$ where an edge is formed between $c_{2,*}$ and $c_{3,*}$ if stable manifold of $c_{2,*}$ is on the boundary of the stable manifold of $c_{3,*}$. The edges are colored depending on if $c_{3,*} \in C_{3,I}$ (red) or $\in C_{3,O}$ (green). The graph is further augmented by the edges within C_2 (blue) if two index 2 saddles' stable manifold have non-empty intersection. For the sake of compactification, we also need to consider the point at infinity which acts as an infinite maximum (m_∞) and therefore is an element of $C_{3,O}$.

We are now equipped with a well-defined structure over the set $C_2 \cup C_3$ which leads to a natural way of clustering the elements in the graph following the hierarchical nature of the stable manifolds. We employ the following three rules to perform the clustering. The rules are applied only on the subsets $C_{2,O}$, $C_{2,S}$ and $C_{3,O} \backslash \{m_\infty\}$.

- **Rule 1**: Two index 2 saddles $c_i, c_j \in C_{2,O}$ are in the same cluster if there is a blue edge between them.
- **Rule 2**: Two maxima $m_i, m_j \in C_{3,O} \backslash \{m_\infty\}$ are in the same cluster if there is a common index 2 saddle c_k which is connected to both m_i and m_j via green edge.
- **Rule 3**: Two index 2 saddles $c_i, c_j \in C_{2,S}$ are clustered together if they each have a green edge to possibly two different maxima $m_i, m_j \in C_{3,O}$ where both m_i, m_j are in the same cluster by Rule 2.

3.3 Detection of Tunnels and Pockets

These three rules produce a clustering of the set $C_{2,O} \cup C_{2,S} \cup C_{3,O} \backslash \{m_\infty\}$. Every cluster is then examined more closely in order to bring out finer invariant features. The index 2 saddles falling in a single cluster can again be of three types as enumerated below.

- Type A: If the stable manifold of an index 2 saddle point is at the boundary of two finite maxima, both from the set $C_{3,O}$.
- Type B: If the stable manifold of an index 2 saddle point is incident upon m_∞ and a single finite maximum from the set $C_{3,O}$.
- Type C: If the stable manifold of an index 2 saddle point is at the boundary of no finite maximum.

The index 2 saddles of type B or type C whose stable manifolds share a common boundary are collected together to form sub-clusters. The combined stable manifold of each such sub-cluster gives a polygonal patch, called *mouth*.

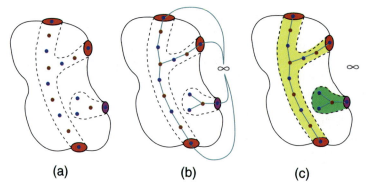

Fig. 3. An illustration of our tunnel and pocket detection algorithm. An imaginary molecular surface is shown with a 3-mouth tunnel and a single pocket. (a) Critical points of h_P are detected. Blue points are index 2 saddles and brown points are maxima. (b) A point at infinity is added and critical points are clustered based on adjacency of stable manifolds. (c) Based on the saddle points incident on infinity, we detect and classify the tunnel (yellow with red mouths) and pocket (green with purple mouth)

The number of *mouths* helps detect the following topological features.

- 0 **Mouth** indicates that the cluster belongs to a *void*.
- 1 **Mouth** indicates that the cluster belongs to a *pocket*.
- $k \geq 2$ **Mouths** indicate the cluster belongs to a *tunnel*. We call it a k-mouthed tunnel.

We use the algorithms described in [21] for computation of the stable manifolds of index 2 saddles. In order to have a computational description of the detected features, we also compute the stable manifolds of the maxima falling into every cluster using the algorithm described in [16]. This produces a tetrahedral decomposition of the features captured. Figure 3 illustrates this process.

3.4 Ranking and Selective Removal of Tunnels and Pockets

The tetrahedral solids describing the pockets and tunnels provide a nice handle to those features and using these handles, the features can be ranked. We primarily use the geometric attributes of the features in order to rank them. Such attributes include, but are not limited to, the combined volume of the tetrahedra and the area of the mouths. The pockets and tunnels are then sorted in order of their increasing geometrically measured importance.

Removal of insignificant features are also made easy because of the volumetric description of the features. As dictated by the applications, a cut-off is set below which the features are considered *noise*. We remove the *topological noise* by marking the outside tetrahedra as inside and updating the surface triangles.

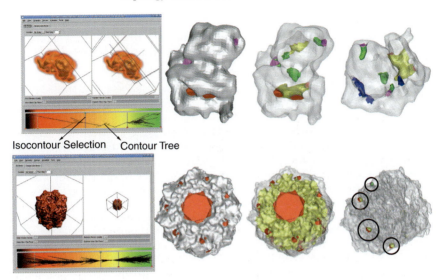

Fig. 4. Results: Top row shows the interface selection for Rieske Iron-sulfur Protein molecule (PDB ID: 1RIE) from a blurred density map. Bottom row shows the isosurface selection for the chaperonin GroEL from cryo-EM density map

4 Results

We show the results of our algorithm on two volumetric data. The top row in Figure 4 shows the electron density volume of Rieske Iron-Sulfur Protein (Protein Data Bank Id: 1RIE). The volume rendering using VolRover [13] is shown in the leftmost subfigure. The tool additionally supports the visualization and isosurface selection using CT. The other subfigures show the selected interface and the detected tunnels and pockets. Note, the mouth of the tunnel is drawn in red and the mouth of the pocket is drawn in purple. The rest of the tunnel surface is drawn in yellow while the pocket surface is drawn in green. The blue patches in the rightmost subfigure shows the filling of the smaller tunnels and pockets. The second row shows the results on the three dimensional scalar volume representing the electron density of the reconstructed image of the chaperonin GroEL from a set of two dimensional electron micrographs. The resolution is $8\mathring{A}$. Using VolRover, a suitable level set is chosen. Note the CT is very noisy and has many branches, because of which it is not possible to extract a single-component isosurface. Nevertheless only one component is vital and the rest of them are merely artifacts caused by noise. The main component along with the detected tunnel is shown next. The result is particularly useful in visualizing the symmetric structure of the chaperonin as depicted in the symmetric set of mouths. In addition to detecting the principal topological feature, the algorithm detects few small tunnels and pockets which are shown separately for visual clarity (rightmost subfigure) and these are removed subsequently as part of the topological noise removal process.

We must also mention that, the presented approach for curation can also be applied to the modeling of smaller subunits of macromolecular complex, like viruses. In such cases, the complex is first segmented from into its building blocks [33]

and they are structurally analyzed either via image processing [34] or via geometry processing [2]. A comprehensive survey on available computational approaches for modeling biological entities from electron density maps can be found in [3].

5 Conclusion

In this paper, we have presented an algorithm which, given an isosurface extracted from a scalar volume, captures the topological and geometric characteristics of the isosurface and allows for the selective removal of unwanted features. The strength of the algorithm lies in its ability to connect the topology of the level set with the critical point structure of the distance function induced by the level set.

Acknowledgment

This research was supported in part by NSF grants IIS-0325550, CNS-0540033 and NIH contracts P20-RR020647, R01-9M074258, R01-GM07308, R01-EB004873.

References

1. Bajaj, C., Bernardini, F., and Xu, G. Automatic reconstruction of surfaces and scalar fields from 3D scans. In *ACM SIGGRAPH* (1995), pp. 109–118.
2. Bajaj, C., and Goswami, S. Automatic fold and structural motif elucidation from 3d EM maps of macromolecules. In *ICVGIP 2006* (2006), pp. 264–275.
3. Bajaj, C., and Yu, Z. Geometric and signal processing of reconstructed 3d maps of molecular complexes: Handbook of computational molecular biology. Computer and Information Sciences Series. Chapman and Hall/CRC press, December 21, 2005.
4. Bajaj, C. L., Pascucci, V., and Schikore, D. R. The contour spectrum. In *Proceeding Visualization '97* (Phoenix, AZ, October 1997), R. Yagel and H. Hagen, Eds., IEEE Computer Society & ACM SIGGRAPH, pp. 167–173.
5. Berman, H. M., Westbrook, J., Feng, Z., Gilliland, G., Bhat, T., Weissig, H., Shindyalov, I., and Bourne, P. The Protein Data Bank. *Nucleic Acids Research* (2000), 235–242.
6. Boyell, R., and Ruston, H. Hybrid techniques for real-time radar simulation. In *Fall Joint Computer Conference '63* (Las Vegas, NV, 1963), pp. 445–458.
7. Carlsson, G., Zomorodian, A., Collins, A., and Guibas, L. J. Persistence barcodes for shapes. *International Journal of Shape Modeling 11*, 2 (2005), 149–187.
8. Carr, H., Snoeyink, J., and Axen, U. Computing contour trees in all dimensions. *Computational Geometry Theory and Applications 24* (2001), 75–94.
9. Carr, H., Snoeyink, J., and van de Panne, M. Simplifying flexible isosurfaces using local geometric measures. *IEEE Visualization 2004* (2004).
10. Chaine, R. A geometric convection approach of 3D reconstruction. In *Proc. Eurographics Sympos. on Geometry Processing* (2003), pp. 218–229.

11. Chazal, F., and Lieutier, A. Stability and homotopy of a subset of the medial axis. In *Proc. 9th ACM Sympos. Solid Modeling and Applications* (2004), pp. 243–248.
12. Collins, A., Zomorodian, A., Carlsson, G., and Guibas, L. A barcode shape descriptor for curve point cloud data. *Computers and Graphics 28* (2004), 881–894.
13. CVC, UT Austin. Volrover. *http://cvcweb.ices.utexas.edu/software/guides.php*.
14. de Berg, M., and van Kreveld, M. Trekking in the alps without freezing or getting tired. *Algorithmica 18*, 3 (1997), 306–323.
15. de Silva, V., and Carlsson, G. Topological estimation using witness complexes. In *Point-Based Graphics '04* (2004), M. Alexa and S. Rusinkiewicz, Eds., pp. 193–199.
16. Dey, T. K., Giesen, J., and Goswami, S. Shape segmentation and matching with flow discretization. In *Proc. Workshop Algorithms Data Strucutres (WADS 03)* (Berlin, Germany, 2003), F. Dehne, J.-R. Sack, and M. Smid, Eds., LNCS 2748, pp. 25–36.
17. Edelsbrunner, H. Surface reconstruction by wrapping finite point sets in space. In *Ricky Pollack and Eli Goodman Festschrift*, B. Aronov, S. Basu, J. Pach, and M. Sharir, Eds. Springer-Verlag, 2002, pp. 379–404.
18. Edelsbrunner, H., Facello, M. A., and Liang, J. On the definition and the construction of pockets in macromolecules. *Discrete Apl. Math. 88* (1998), 83–102.
19. Edelsbrunner, H., Letscher, D., and Zomorodian, A. Topological persistence and simplification. *Discrete Comput. Geom. 28* (2002), 511–533.
20. El-Sana, J., and Varshney, A. Controlled simplification of genus for polygonal models. In *Proceedings of the IEEE Visualization '97* (1997), pp. 403–412.
21. Giesen, J., and John, M. The flow complex: a data structure for geometric modeling. In *Proc. 14th ACM-SIAM Sympos. Discrete Algorithms* (2003), pp. 285–294.
22. Goswami, S., Dey, T. K., and Bajaj, C. L. Identifying flat and tubular regions of a shape by unstable manifolds. In *Proc. 11th ACM Sympos. Solid and Phys. Modeling* (2006), pp. 27–37.
23. Goswami, S., Gillette, A., and Bajaj, C. Efficient Delaunay mesh generation from sampled scalar function. In *Proc. 16th Int. Meshing Roundtable* (2007), p. to appear.
24. Guillemin, V., and Pollack, A. *Differential Topology*. Prentice-Hall Inc., Englewood Cliffs, New Jersey, 1974.
25. Guskov, I., and Wood, Z. Topological noise removal. In *Graphics Interface* (2001), pp. 19–26.
26. Ju, T., Losasso, F., Schaefer, S., and Warren, J. Dual contouring of hermite data. In *Proceedings of SIGGRAPH* (2002), pp. 339–346.
27. Lorensen, W., and Cline, H. Marching cubes: A high resolution 3d surface construction algorithm. In *ACM SIGGRAPH '87* (1987), pp. 163–169.
28. Nooruddin, F., and Turk, G. Simplification and repair of polygonal models using volumetric techniques. Tech. Rep. Research Report 99-37, Georgia Tech., 1999.
29. Pascucci, V., and Cole-McLaughlin, K. Parallel computation of the topology of level sets. *Algorithmica 38* (2003), 249–268.
30. Siersma, D. Voronoi diagrams and morse theory of the distance function. In *Geometry in Present Day Science*, O. E. Barndorff and E. B. V. Jensen, Eds. 1999, pp. 187–208.

31. van Kreveld, M., van Oostrum, R., Bajaj, C., Pascucci, V., and Schikore, D. Contour trees and small seed sets for isosurface traversal. In *Proc. 13th ACM Symposium on Computational Geometry* (1997), pp. 212–220.
32. Wood, Z., Hoppe, H., Desbrun, M., and Schroder, P. Removing excess topology from isosurfaces. *ACM Transactions on Graphics 23*, 2 (April 2004), 190–208.
33. Yu, Z., and Bajaj, C. Automatic ultrastructure segmentation of reconstructed cryoem maps of icosahedral viruses. *IEEE Transactions on Image Processing: Special Issue on Molecular and Cellular Bioimaging 14*, 9 (2005), 1324–37.
34. Yu, Z., and Bajaj, C. Computational approaches for automatic structural analysis of large bio-molecular complexes. *IEEE/ACM Transactions on Computational Biology and Bioinformatics accepted for publication* (2007).
35. Zhang, X., Bajaj, C., Kwon, B., Dolinsky, T., Nielsen, J., and Baker, N. Application of new multiresolution methods for the comparison of biomolecular electrostatics properties in the absence of structural similarity. *Multiscale Modeling and Simulation 5*, 4 (2006), 1196–1213.
36. Zomorodian, A. *Computing and Comprehending Topology: Persistence and Hierarchical Morse Complexes*. PhD thesis, University of Illinois at Urbana-Champaign, 2001.
37. Zomorodian, A., and Carlsson, G. Computing persistent homology. *Discrete Comput. Geom. 33*, 2 (2005), 249–274.

Representing Interpolant Topology for Contour Tree Computation

Hamish Carr[1] and Jack Snoeyink[2]

[1] University College Dublin hamish.carr@ucd.ie
[2] University of North Carolina at Chapel Hill snoeyink@cs.unc.edu

Summary. Algorithms for computing contour trees for visualization commonly assume that the input is defined by barycentric interpolation on simplicial meshes or by trilinear interpolation on cubic meshes. In this paper, we describe a general framework for computing contour trees from a graph that captures all significant topological features. We show how to construct these graphs from any mesh-based interpolant by using cell-by-cell "widgets," and also how to avoid constructing the entire graphs by making finite state machines that capture their traversals.

Our framework eases algorithm development and implementation, and can be used to establish relationships between interpolants. For example, we use it to demonstrate a formal equivalence between the topology defined by implicitly disambiguated marching cubes cases and the topology induced by 8-/18- digital image connectivity.

1 Introduction

In a scalar field, a *level set* is the set of points with a specified *isovalue*: the connected components of the isosurface are called *contours*. In three-dimensional scalar fields, level sets are known as *isosurfaces*: isosurfaces and contours are fundamental for segmenting and rendering these scalar fields. The nesting relationships of contours for all possible isovalues is expressed by a diagram called the *contour tree*, which has been used for isosurface extraction [31, 5], for abstract representation of the field [1, 24, 7], to index individual contours [3, 14, 5] and their geometric properties [7], to guide simplification of input meshes [10], and to compare scalar fields [33].

Although recent algorithms for computing the contour tree are efficient, they were originally defined only for simplicial meshes [1, 6], then extended to trilinear meshes [24] and digital images [18]. In this paper, we give a firm theoretical base, not only for these extensions, but for extension to any arbitrary mesh-based interpolant, tessellation kernel, or connectivity rule. We also simplify methods for computing contour trees on bilinear, trilinear, and quadratic

Bézier interpolants, Marching Cubes [16, 19], Marching Hypercubes [2] and digital image connectivity.

Our motivation is that topological computations should be possible for any interpolant chosen, since no interpolant is ideal for all cases.

For some purposes, users will want high order interpolants. For example, numerical simulations often require higher-order interpolants to obtain high-order continuity—the trilinear interpolant may not be enough because it is only C^0 continuous.

For other purposes, low order interpolants will suffice. Generating iso-surfaces from a trilinear interpolant is already not easy, as many works between [16] and [15, 21] demonstrate. Furthermore, contour tree algorithms depend on sorting critical points, so numerical error, e.g., in positioning the body saddle of a degree-3 polynomial, can result in incorrect topology. Quantization and noise of sampled volumetric data can create spurious features below the scales of interest. Thus, practical methods like implicitly disambiguated Marching Cubes cases (e.g., from [19]) or segmentation algorithms with low-order digital connectivity remain popular for their simplicity, speed and robustness.

In this paper, we describe a framework which can adapted to many interpolants and applied to any of the known contour tree algorithms. We do so by extending the "oracle" of Pascucci & Cole-McLaughlin [24]. Section 2 gives a brief overview of the contour tree, its nesting properties and existing algorithms for computing it, focussing on the graph-theoretic nature of the algorithms. Section 3 then shows how to extend these algorithms by characterizing the properties of *topology graphs* sufficient for the correct computation of the contour tree, using the bilinear interpolant as a simple example. Section 4 shows some simple repeating graph structures, or *widgets*, that can be used to construct topology graphs for bilinear and trilinear interpolants, Section 5 then shows how to represent the overall topology of an interpolant as a finite state machine whose states correspond to the individual tessellation cases for isosurface extraction, and how to use these finite state machines to define suitable graphs to compute the contour tree. Section 6 then extends this approach to the Marching Cube cases of Montani et al. [19] and to Marching Hypercubes [2], then shows that the topology of Marching Hypercubes is equivalent to standard digital image connectivity rules. Section 7 gives some comments on implementation, while Section 8 gives our conclusions, and future extensions.

2 The Contour Tree

In a scalar field $f\colon \mathbb{R}^3 \to \mathbb{R}$, the *level set* of an *isovalue* h is the set $L(h) = \{(x, y, z) \mid f(x, y, z) = h\}$. A *contour* is a connected component of a level set. As h increases, contours appear at local minima, join or split at saddles, and disappear at local maxima. Shrinking each contour to a point gives the *contour tree* tracking this evolution: a tree because the domain $\mathbb{R}^3$ is simply-connected. In more general domains it is the *Reeb graph* [25, 26], which is used to study manifold topology.

Data is not always a continuous function f: It may be represented by a mesh M and a procedure that tessellates isosurfaces by cases that do not correspond exactly to any mathematical interpolant. Or, as in digital images, it may be explicitly discontinuous between samples that are implicitly connected according to a known rule.

An older definition of the contour tree that applies also to non-continuous representations is as the nesting relationship of a set of known contours [3]. We must be careful with boundary conditions, since *nesting* imports ideas of enclosure, but most practical difficulties vanish if we consider nesting relationship of connected subsets of the form $\{x : f(x) \geq h\}$ or $\{x : f(x) \leq h\}$. An example that we develop in Section 6 is the use of 4/8 connectivity in digital imaging.

2.1 Previous Work

The contour tree has been used to index and extract contours [3, 31, 5], to describe terrain [13, 27] or volumetric data [26, 1], to detect features [33, 28], to simplify data [10, 7], to design transfer functions [28, 32] and to extract contour properties [14, 7].

Known algorithms compute the contour tree for 4/8 connectivity [27], 8/26 connectivity [28, 18], simplicial meshes [31, 29, 6, 9], and trilinear interpolants [24]. We unify and extend all of these algorithms to arbitrary meshes in any dimension and to surface tessellators such as Marching Cubes [16, 19] or Marching Hypercubes [2] that are procedural, rather than derived from an underlying interpolation scheme.

Takeshima et al. [27] computed contour trees in 2 dimensions from *surface networks* of monotone paths between critical points. Later extended to three dimensions [28], this algorithm runs in $O(n^2)$. Special treatment was required for boundary cases and for multiple saddles. Chiang et al. [9] improved this by using the surface network as input for the algorithm of Carr, Snoeyink & Axen [6].

Van Kreveld et al. [31] swept an explicit polygonal contour through a simplicial mesh in each isovalue direction then used the contour tree to accelerate contour extraction. Extended by Tarasov & Vyalyi [29] and Pascucci [23] this computes the contour tree in $O(N \log(N))$ steps in any dimension, including tracking topological genus. Multiple saddles and boundary cases required special handling.

Carr, Snoeyink & Axen [6] achieved $O(N + n\log(n) + t\alpha(t))$ time for a simplicial mesh M in any dimension with N cells, n vertices, and t critical points. Each vertex is assigned a value, and the function f is obtained by piecewise linear (barycentric) interpolation over each simplex. Pascucci & Cole-McLaughlin [24] parallelized this, modelled trilinear interpolants, and added topological genus, but did not handle arbitrary interpolants specifically. This algorithm has two basic phases: join/split tree computation and tree merging.

The first phase computes the *join tree* and *split tree* for a simplicial mesh. Since these are symmetric in the isovalue, we consider only the split tree. Where the contour tree records the connectivity of $\{x : f(x) = h\}$, the split tree records the connectivity of $\{x : f(x) \geq h\}$. This is demonstrably equivalent to the connectivity of $\{v \in G : f(v) \geq h\}$, where G is the graph of the vertices and edges of the mesh M (it's one-skeleton). This equivalence follows from three properties of the induced graph G:

I. Edges of G represent paths in the domain that are monotone in f.
II. G contains all splits and local maxima of the function f.
III. For any h, two vertices $u, v \in G$ are connected above h iff u, v are connected in the mesh by a path above h.

Based on this equivalence, the algorithm adds vertices and edges to a union-find structure [30] in decreasing order, maintaining the connected sets of $\{\sigma \in M : f(\sigma) \geq h\}$ incrementally for decreasing h. Local maxima are detected as having no neighbors with higher-valued vertices, and splits are detected as having neighbors in separate components, which are then unioned. It is easy to construct a tree on local maxima, splits, and the global minimum that records these changes.

The contour tree is a merge of the join and split trees, using a couple of invariants: Every edge leading upwards from a leaf in the join tree always appears in the contour tree. Similarly, every edge leading downwards from a leaf in the split tree always appears in the contour tree. Moreover, this is recursively true for the trees left over after the leaf vertex has been removed from all three trees. Thus, the contour tree can be computed by repeatedly choosing a suitable leaf vertex in the join or split trees, adding its edge to the contour tree, and deleting it from join and split trees.

Pascucci & Cole-McLaughlin [24] also compute topological genus changes by adding to the three trees all the *Morse critical points [17]*, which are the points at which local connectivity changes. These points are determined numerically.

We will see shortly that the contour tree construction algorithm can be extended to an arbitrary mesh by choosing any graph G that satisfies properties I and II above. This idea is already implicit in the divide-and-conquer computation of the contour tree for the trilinear interpolant by Pascucci & Cole-McLaughlin [24]. They captured the topology of the trilinear interpolant by separately computing the join and split trees for subregions, and using the unions of these trees as graph input to compute join, split and contour trees for larger and larger regions. We extend this contour tree computation to arbitrary meshes by varying the graph input further, and by computing oracles directly from isosurface topology cases for the desired interpolant.

3 Join and Split Graphs

As we saw above, the contour tree computation for a simplicial mesh M exploits the fact that the induced graph G satisfies I-III of the following for split trees and I, IV and V for join trees.

I. Edges of G represent paths in the domain that are monotone in f.
II. G contains all splits and local maxima of the function f.
III. For any h, two vertices $u, v \in G$ are connected above h iff u, v are connected in the mesh by a path above h.
IV. G contains all joins and local minima of the function f.
V. For any h, two vertices $u, v \in G$ are connected below h iff u, v are connected in the mesh by a path below h.

To extend contour tree computation to arbitrary interpolants, we can use any graph that satisfies these properties. We use *split graph* to mean a graph that satisfies I-III, and *join graph* to mean a graph that satisfies I, IV & V. When genus information is also to be computed, we add all Morse critical points to these graphs.

Global critical points (the minima, maxima, and saddles of the function) that lie on interiors of cells must be critical points of cells to which they belong; those that lie on boundaries of cells must be critical points of the intersections between cells. Any graph that includes all local maxima and splits within cells and their boundaries therefore satisfies II, while any graph that contains all local minima or joins satisfies IV.

Any of the following graphs satisfies III: the contour tree, the Morse-Smale Complex [12], the surface network [27], simplicial mesh edges, the union of cellwise contour trees or split trees [24] or the union of cellwise split graphs. In fact, our definition of split graphs is complementary to the "oracle" of Pascucci & Cole-McLaughlin [24] as split graphs can be used by the oracle to compute cellwise split trees. In the following sections, we will show how to construct split graphs with *widgets*, which are graphs that capture all possible topological changes in a cell for a given interpolant, or with finite state machines that track topological changes in the contours.

4 Graph Widgets for Standard Interpolants

For the two most common non-simplicial interpolants, bilinear in the plane and trilinear in 3d, we need not explicitly construct a graph that satisfies properties I-V; it is enough to define a standard small *graph widget*, a graph that applies to all cells, and a procedure for each cell to determine the values assigned to vertices in this graph. The graphs of the widgets for bilinear and trilinear interpolants tell the join and split tree algorithm which unions and finds to perform; the values assigned at vertices tell the order. Using one graph for all cells saves memory and processing time.

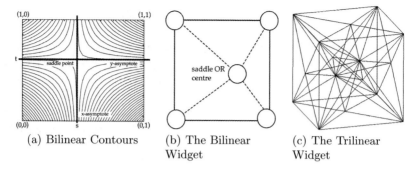

(a) Bilinear Contours (b) The Bilinear Widget (c) The Trilinear Widget

Fig. 1. Widget Construction for Bilinear & Trilinear Interpolants. Where a bilinear cell includes the saddle point, the saddle point is inserted with the edges marked. Where no saddle point exists, the centre point will be a regular point, so this remains a split graph. For the trilinear cell, each face is a bilinear widget with the value of the face saddle or center, and two body saddles are inserted and connected to all face saddles and all corner vertices but one by monotone paths

As shown in Figure 1, bilinear contours are families of hyperbolae with vertical and horizontal asymptotes passing through a saddle point [22]. For cells that include the saddle, therefore, the saddle point is a split and must be included in the split graph. In addition, edges from the saddle point to each vertex are necessary to satisfy III & V. For cells that do not include the saddle, we can still include any point within the cell and edges to the vertices without violating any of the properties. As a result, the graph widget shown in Figure 1(b) is a sufficient join and split graph for the cell.

The value at the added vertex can be computed by observing where partial derivatives of the interpolant vanish. Using f_{xy} to denote the value of the original vertex (x,y) of the unit square $[0,1] \times [0,1]$, the saddle point is outside this cell if the signs of $(f_{11} - f_{01})$ and $(f_{00} - f_{10})$ are opposite or the signs of $(f_{11} - f_{10})$ and $(f_{00} - f_{01})$ are opposite. If the saddle is in the cell or on the boundary, the value of the new vertex is the saddle: $(f_{00}f_{11} - f_{01}f_{10})/(f_{00} - f_{01} + f_{11} - f_{10})$. Otherwise, we can simply average the values at the corners.

The graph widget of Figure 1(c) can be used for the trilinear interpolant: For a cell, we build bilinear widgets in each face and add two vertices for the cell body. From the quadratic equation given by Natarajan [20], there are at most two body saddles [21], which are aligned along a major diagonal of the cell. Each body saddle is connected to each face saddle and to each vertex except the one occluded along the major diagonal by the other body saddle— that is the only vertex that cannot be reached from the body saddle by a path in the cell that is monotone in f. Where only one body saddle is in the cell, the two body saddles may be set to the same point, and where no body saddles occur, the two body saddles are set to the centre or a corner of the cell. This solution has been implemented [32] and gives the same topology as Pascucci & Cole-McLaughlin [24].

5 Finite State Machines for Split Graphs

Because the algorithms to compute join and split trees essentially sweep the underlying graph or graph widgets, we can reduce the computation further to a process of generating the correct unions and finds by a finite state machine. In fact, we can also generalize from interpolants with graph widgets, to any method (interpolant or set of isosurface tessellation cases) that produces contours that nest properly and move continuously across cell boundaries. As long as there is an underlying graph that satisfies properties I-V, we can rest assured that our contour tree is correctly captures the nesting.

Consider a marching algorithm that extracts contours in each cell. Such an algorithm is *accurate* for a given interpolant if the contours extracted for any isovalue are homotopic to the contours of the interpolant in the cell. Two such extracted contours are *equivalent* if they are homotopic to each other and partition the vertices of the cell in the same way. It follows that the tessellation cases of the algorithm are the equivalence classes of contours it extracts.

Define vertices and Morse critical points of a cell to be *potential critical points*: *potential* because they may not be global critical points. However, all global critical points must be potential critical points for at least one cell. Moreover, all topological changes occur at potential critical points: Morse critical points because the topology of the contours change, and vertices because the vertex partition changes.

To extract a split graph, we track the vertex partition of that cell as an isovalue sweep passes each potential critical point. At each potential critical point v, we add at least one edge vw to the split graph for each component that merges at v. Each vertex w may be any potential critical point in the component: choosing the lowest valued one means that the resulting split graph is the split tree. However, we assume the general case: split graph rather than split tree.

Instead of computing split graphs in advance for each cell, we observe that there are a finite number of tessellation cases for each algorithm, and that the sweep always increases the number of potential critical points that have been passed. It therefore follows that there are at most N_c^2 possible transitions for N_c tessellation cases. Moreover, each transition corresponds exactly to a sweep past a particular potential critical point.

We describe these transitions as a Mealy machine – a finite state machine with outputs on transitions: $FSM = \{\Sigma, \Lambda, Q, q_{init}, F, \delta\}$, where Σ, the input alphabet, consists of the potential critical points, Λ, the output alphabet, consists of edges for the split graph, Q, the set of states, consists of the tessellation cases, q_{init}, the initial state, is the case with no potential critical points above the isosurface, F, the set of terminal states, includes only the case with all potential critical points above, δ, the transition function, permits transitions for any potential critical point below the isosurface and outputs the edges to be added to the split graph for the cell to represent topological differences between the states before and after the transition.

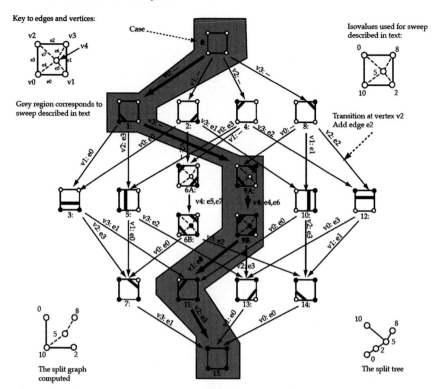

Fig. 2. Finite State Machine for a Bilinear Split Graph: Figures with states illustrate the tessellation cases; transitions are annotated with the input vertex that is passed and output edge added to the split graph. The grey region indicates the path of a sweep through the single cell shown in the top right corner

Figure 2 shows the bilinear split graph finite state machine: the grey region represents a sweep through the cell in the top right corner, progressing from state 0 through states 1, 9A, 9B, 11 and 15, extracting the split graph shown in the bottom left corner: the corresponding split tree is shown in the bottom right corner.

Since this sweep is identical to the sweep through the function used to compute the split tree, a side benefit of this is that the oracle can be discarded. Instead of using an oracle to determine a split graph for each cell, we need only store the connected components (i.e. state) associated with each cell during the sweep. For each potential critical point in the entire mesh, we update the state of all cells to which the potential critical point belongs, and add the corresponding edges to the split graph for the cell or directly to the union-find structure in the contour tree algorithm's split tree sweep.

Figure 3 shows an overview of the finite state machine for split graphs of the trilinear interpolant; a larger PDF-format version that can be explored

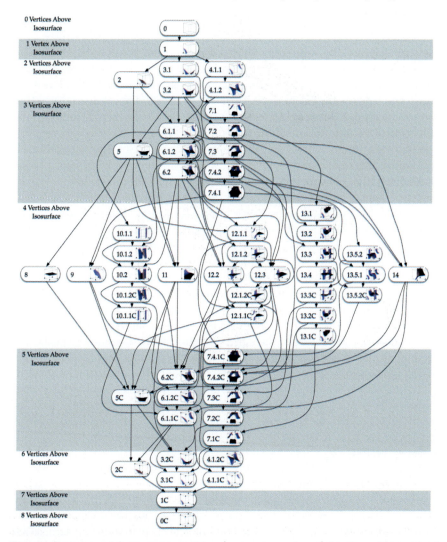

Fig. 3. Overview of finite state machine (symmetry-reduced) for the split graph of a trilinear interpolant; link to larger, on-line version in text. States display an isosurface, and are numbered first according to the Marching Cubes cases of Montani, Scateni, & Scopigno [19], which depends on the pattern of vertices above/below the isosurface, then these are divided by whether face and body saddles are above or below. Transitions occur as the isosurface passes the value of a vertex, face saddle, or body saddle, and produce edges that are added to the split graph; transition labels have been omitted from the diagram to avoid futher clutter. Every possible evolution of isosurface topology is reflected in a directed path through this graph

in detail is available on-line at http://www.cs.unc.edu/~snoeyink/trilinearFSM.pdf. This finite state machine was constructed directly from the trilinear cases by examining which vertices could possibly be swept past in each case. Cases are numbered according to the Marching Cubes cases of Montani, Scateni, & Scopigno [19], which depends on the pattern of vertices above and below a canonical value, then subdivided by the connectivity within the cell body and faces. To conserve space, we reduce cases by symmetry and omit transition annotations from the edges.

6 Marching (Hyper)Cubes and Digital Images

Given that we build our finite state machines directly from the tessellation cases, it is natural to ask whether similar finite state machines exist for tessellation cases that nest but are not accurate for any specific interpolant, e.g. the original Marching Cubes [16]. Because of the well-known cracking problem [11], the original cases are not continuous across cell boundaries, so we instead use the cases of Montani, Scateni, & Scopigno [19] in the split finite state machine shown in Figure 4, which is simply the result of collapsing subcases of our finite state machine for the trilinear interpolant.

In general, provided that the surfaces extracted nest inside each other, or are homotopic to a set of surfaces that do so, these finite state machines represent the topology of any set of tessellations. Since the surfaces generated by Marching Cubes are linear triangulations based on linearly interpolated edge points, it is easy to see that each case sweeps out a range of surfaces over the set of isovalues that use that case. Moreover, since the cases enclose an increasing set of vertices, it is also true that the sequential cases are (topologically at least) nested.

The more general Marching Hypercubes cases of Bhaniramka, Wenger & Crawfis [2] define n-dimensional tessellation cases as the boundary of the convex hulls of the set of all black vertices along with any points interpolated along the edges. Since these sets of points get larger as the isovalue decreases, and the convex hulls also expand outwards, it is clear that these cases satisfy the nesting property required for the contour tree to be computable.

Moreover, we note that the effect of the Marching Hypercubes rule is that any two black vertices in the cell are always enclosed within the same surface, implying that when a given vertex is swept past (i.e. a transition in our finite state machine occurs), it is connected not only to its black edge neighbours but also to its black face- and body- diagonal neighbours. In comparison, during the reverse sweep, white vertices are only connected to white edge-connected neighbours.

These rules can be implemented for the split graph by connecting each vertex to every other black vertex, and in the join graph by connecting each vertex only to white edge neighbours, thus dispensing with the oracle and finite state machine entirely, replacing them with a simple graph widget in every cell.

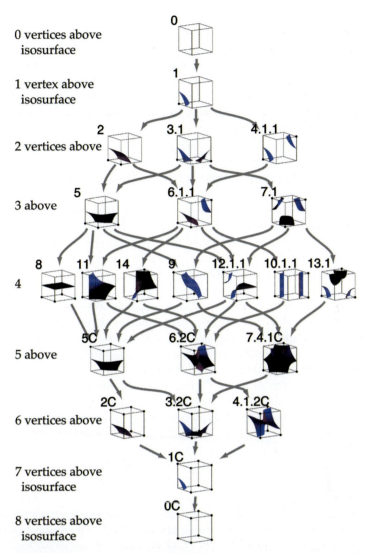

Fig. 4. Finite State Machine (Marching Cubes [19]) (Symmetry-Reduced). Marching Cubes selects a subset of the cases identified for the trilinear interpolant, selects and collapses the transitions between these cases. Topological consistency can be checked along all directed paths. E.g., each path must be extendable from state 0 to 0C. Where the trilinear interpolant may have had a sequence of transitions on face and body saddles that output a sequence of edges for the split graph, Marching Cubes has a single transition that would output these edges all at once

We note that since the downwards sweep connects every vertex to each of its edge-, face- and body- neighbours, and each upwards sweep connects to edge neighbours only, this widget is equivalent to the 26-/8- connectivity rule used to compute foreground/background connectivity in digital images. And, if we reverse the Marching Hypercubes rule to use the convex hull of the white vertices, we will get the more commonly used 8-/26- connectivity rule in three dimensions, 4-/8- connectivity in two dimensions, and $2^d - /(3^d - 1)-$ connectivity rule in d dimensions.

Working backwards from this observation, we now see that the Marching Cubes cases of Montani, Scateni & Scopigno [19] disambiguate faces by assuming that black vertices are never face- or body- connected, while white vertices are face- but not body- connected. This is equivalent to 8-/18- digital connectivity, and can be converted to 8-/26- connectivity by substituting two cases (4.1.2C for 4.1.1C).

7 Implementation and Results

We have identified two methods of computing suitable graphs for contour tree computation, widgets and finite state machines. We find that widgets are easier to implement than the more general finite state machines. We also find that building and displaying a finite state machine generally leads to a suitable widget that can be used in their place.

The trilinear widget shown in Figure 1 has been implemented successfully by Weber et al. [32], and the 8-/18- rule was used for flexible isosurfaces [5].

Table 1 shows construction times for the contour tree using the minimal 5-fold subdivision scheme described in [4] and using our Marching Cubes method. The contour tree sizes in this table are before removing perturbation, but are similar since large scale topology is unaffected by local decisions. Computation speed is similar, since the two methods process approximately the same number of edges.

Using Marching Cubes, however, results in a clear gain in rendering speed, as shown in Table 2. We took the minimum and maximum sampled isovalues,

Table 1. Comparison of Construction Times. Broadly speaking, Marching Cubes topology is computed as rapidly as simplicial topology

File	Data Size	Five Simplices			Marching Cubes		
		Join/Split	Merge	Tree Size	Join/Split	Merge	Tree Size
3dknee	$127 \times 256 \times 256$	141.34s	32.28s	2,751,506	129.70s	31.12s	2,706,019
3dhead	$109 \times 256 \times 256$	89.05s	23.94s	2,231,900	81.62s	22.86s	2,196,594
1dog.0.8	$72 \times 64 \times 60$	1.63s	0.19s	18,498	1.61s	0.18s	17,656
3gap.0.8	$29 \times 60 \times 131$	1.31s	0.15s	14,290	1.28s	0.15s	13,164
neghip	$64 \times 64 \times 64$	1.00s	0.08s	2,544	0.99s	0.08s	2,063
fuel	$64 \times 64 \times 64$	0.91s	0.06s	299	0.94s	0.06s	227

Table 2. Sample Isosurface Sizes for Simplicial Subdivision and Marching Cubes. Marching Cubes outperforms simplicial subdivision by a factor of about 2.3, roughly in accordance with the estimates in [8], while avoiding the directional biases in [4]

File	Size	Isovalue	Five Simplices	Marching Cubes	Ratio
3dknee	$127 \times 256 \times 256$	1639.2	3,662,308	1,634,744	2.24
3dhead	$109 \times 256 \times 256$	1639.2	1,043,892	441,998	2.36
1dog.0.8	$72 \times 64 \times 60$	0.30	393,450	165,964	2.37
3gap.0.8	$29 \times 60 \times 131$	0.65	194,568	81,798	2.38
neghip	$64 \times 64 \times 64$	101.9	49,484	20,360	2.43
fuel	$64 \times 64 \times 64$	101.9	7,564	2,946	2.57

and arbitrarily chose an isovalue at 40% of this range. We then extracted the isosurface using the minimal (five simplices) subdivision and the Marching Cubes implementations. In each case, the simplicial subdivision resulted in roughly two to two-and-a-half times as many triangles, which we expected from the results of Carr, Theußl and Möller [8]. Moreover, we know from [4] that simplicial subdivision produces isosurfaces with visible directional biases, where Marching Cubes does not.

Marching Cubes is both faster and better quality than simplicial subdivision, which is the main reason why we have used it. In practice, the bottleneck for exploratory visualization is triangle rendering, which dominates almost any other cost. Thus, using Marching Cubes rather than simplices is a major advantage in practice.

8 Conclusions and Future Work

In this paper, we have shown how to extend contour tree computation from simplices and trilinear interpolants to arbitrary interpolants, to non-interpolating tessellation cases such as Marching Hypercubes and to digital images, while keeping execution times at least as fast as for simplicial meshes when using Marching Cubes.

We have also demonstrated a finite state machine model that tracks the topology of a given interpolant, aiding in understanding the evolution of isosurfaces in a single cell, and described simple graph widgets for bilinear and trilinear interpolants that can be used instead of the oracle of Pascucci & Cole-McLaughlin.

In the future, we intend to build finite state machines explicitly and compare their performance with the simple widgets shown above, and to investigate similar widget and finite state machine representations of gradient and flow interpolation.

Acknowledgements

Acknowledgements are due to the University of British Columbia, to the National Science and Engineering Research Council (Canada), to the University of North Carolina at Chapel Hill, to the National Science Foundation (US), to University College Dublin and to Science Foundation Ireland for supporting this work with fellowships, research grants and equipment.

References

1. C. L. Bajaj, V. Pascucci, and D. R. Schikore. The Contour Spectrum. In *Proceedings of Visualization 1997*, pages 167–173, 1997.
2. P. Bhaniramka, R. Wenger, and R. A. Crawfis. Isosurface Construction in Any Dimension Using Convex Hulls. *IEEE Transactions on Visualization and Computer Graphics*, 10(2):130–141, 2004.
3. R. L. Boyell and H. Ruston. Hybrid Techniques for Real-time Radar Simulation. In *Proceedings of the 1963 Fall Joint Computer Conference*, pages 445–458. IEEE, 1963.
4. H. Carr, T. Möller, and J. Snoeyink. Simplicial Subdivisions and Sampling Artifacts. In *Proceedings of Visualization 2001*, pages 99–106, 2001.
5. H. Carr and J. Snoeyink. Path Seeds and Flexible Isosurfaces: Using Topology for Exploratory Visualization. In *Proceedings of Eurographics Visualization Symposium 2003*, pages 49–58, 285, 2003.
6. H. Carr, J. Snoeyink, and U. Axen. Computing Contour Trees in All Dimensions. *Computational Geometry: Theory and Applications*, 24(2):75–94, 2003.
7. H. Carr, J. Snoeyink, and M. van de Panne. Simplifying Flexible Isosurfaces with Local Geometric Measures. In *Proceedings of Visualization 2004*, pages 497–504, 2004.
8. H. Carr, T. Theußl, and T. Möller. Isosurfaces on Optimal Regular Samples. In *Proceedings of Eurographics Visualization Symposium 2003*, pages 39–48, 284, 2003.
9. Y.-J. Chiang, T. Lenz, X. Lu, and G. Rote. Simple and Optimal Output-Sensitive Construction of Contour Trees Using Monotone Paths. *Computational Geometry: Theory and Applications*, 30:165–195, 2005.
10. Y.-J. Chiang and X. Lu. Progressive Simplification of Tetrahedral Meshes Preserving All Isosurface Topologies. *Computer Graphics Forum*, 22(3), 2003.
11. M. Dürst. Letters: Additional Reference to "Marching Cubes". *Computer Graphics*, 22(4):65–74, 1988.
12. H. Edelsbrunner, J. Harer, and A. Zomorodian. Hierarchical Morse Complexes for Piecewise Linear 2-Manifolds. In *Proceedings of the 17th ACM Symposium on Computational Geometry*, pages 70–79. ACM, 2001.
13. H. Freeman and S. Morse. On Searching A Contour Map for a Given Terrain Elevation Profile. *Journal of the Franklin Institute*, 284(1):1–25, 1967.
14. L. Kettner, J. Rossignac, and J. Snoeyink. The Safari Interface for Visualizing Time-Dependent Volume Data Using Iso-surfaces and Contour Spectra. *Computational Geometry: Theory and Applications*, 25(1–2):97–116, 2001.

15. A. Lopes and K. Brodlie. Improving the robustness and accuracy of the marching cubes algorithm for isosurfacing. *IEEE Transactions on Visualization and Computer Graphics*, 9(1):16–29, 2003.
16. W. E. Lorenson and H. E. Cline. Marching Cubes: A High Resolution 3D Surface Construction Algorithm. *Computer Graphics*, 21(4):163–169, 1987.
17. J. Milnor. *Morse Theory.* Princeton University Press, Princeton, NJ, 1963.
18. S. Mizuta and T. Matsuda. Description of the Topological Structure of Digital Images by Region-based Contour Tree and Its Application. Technical report, Institute of Electronics, Information and Communication Engineers, 2004.
19. C. Montani, R. Scateni, and R. Scopigno. A modified look-up table for implicit disambiguation of Marching Cubes. *Visual Computer*, 10:353–355, 1994.
20. B. Natarajan. On generating topologically consistent isosurfaces from uniform samples. *Visual Computer*, 11:52–62, 1994.
21. G. M. Nielson. On Marching Cubes. *IEEE Transactions on Visualization and Computer Graphics*, 9(3):283–297, 2003.
22. G. M. Nielson and B. Hamann. The Asymptotic Decider: Resolving the Ambiguity in Marching Cubes. In *Proceedings of Visualization 1991*, pages 83–91. IEEE, 1991.
23. V. Pascucci. On the Topology of the Level Sets of a Scalar Field. In *Abstracts of the 13th Canadian Conference on Computational Geometry*, pages 141–144, 2001.
24. V. Pascucci and K. Cole-McLaughlin. Parallel Computation of the Topology of Level Sets. *Algorithmica*, 38(2):249–268, 2003.
25. G. Reeb. Sur les points singuliers d'une forme de Pfaff complètement intégrable ou d'une fonction numérique. *Comptes Rendus de l'Académie des Sciences de Paris*, 222:847–849, 1946.
26. Y. Shinagawa, T. L. Kunii, and Y. L. Kergosien. Surface Coding Based on Morse Theory. *IEEE Computer Graphics and Applications*, 11:66–78, September 1991.
27. S. Takahashi, T. Ikeda, Y. Shinagawa, T. L. Kunii, and M. Ueda. Algorithms for Extracting Correct Critical Points and Constructing Topological Graphs from Discrete Geographical Elevation Data. *Computer Graphics Forum*, 14(3):C-181–C-192, 1995.
28. S. Takahashi, Y. Takeshima, and I. Fujishiro. Topological volume skeletonization and its application to transfer function design. *Graphical Models*, 66(1):24–49, 2004.
29. S. P. Tarasov and M. N. Vyalyi. Construction of Contour Trees in 3D in $O(n \log n)$ steps. In *Proceedings of the 14th ACM Symposium on Computational Geometry*, pages 68–75, 1998.
30. R. E. Tarjan. Efficiency of a good but not linear set union algorithm. *Journal of the ACM*, 22:215–225, 1975.
31. M. van Kreveld, R. van Oostrum, C. L. Bajaj, V. Pascucci, and D. R. Schikore. Contour Trees and Small Seed Sets for Isosurface Traversal. In *Proceedings of the 13th ACM Symposium on Computational Geometry*, pages 212–220, 1997.
32. G. Weber, S. Dillard, H. Carr, V. Pascucci, and B. Hamann. Topology-based, flexible volume rendering. To appear in IEEE Transactions on Visualization and Computer Graphics, 2007.
33. X. Zhang, C. L. Bajaj, and N. Baker. Fast Matching of Volumetric Functions Using Multi-resolution Dual Contour Trees. Technical report, Texas Institute for Computational and Applied Mathematics, Austin, Texas, 2004.

Path Line Attributes - an Information Visualization Approach to Analyzing the Dynamic Behavior of 3D Time-Dependent Flow Fields

Kuangyu Shi[1], Holger Theisel[2], Helwig Hauser[3], Tino Weinkauf[4], Kresimir Matkovic[3], Hans-Christian Hege[4], and Hans-Peter Seidel[1]

[1] MPI Informatik, 66123 Saarbrücken, Germany
 {skyshi,hpseidel}@mpi-inf.mpg.de
[2] Bielefeld University, 33501 Bielefeld, Germany
 theisel@techfak.uni-bielefeld.de
[3] VRVis Vienna, 1220 Vienna, Austria {hauser,matkovic}@vrvis.at
[4] Zuse Institute Berlin, 14159 Berlin, Germany {weinkauf,hege}@zib.de

Summary. We describe an approach to visually analyzing the dynamic behavior of 3D time-dependent flow fields by considering the behavior of the path lines. At selected positions in the 4D space-time domain, we compute a number of local and global properties of path lines describing relevant features of them. The resulting multivariate data set is analyzed by applying state-of-the-art information visualization approaches in the sense of a set of linked views (scatter plots, parallel coordinates, etc.) with interactive brushing and focus+context visualization. The selected path lines with certain properties are integrated and visualized as colored 3D curves. This approach allows an interactive exploration of intricate 4D flow structures. We apply our method to a number of flow data sets and describe how path line attributes are used for describing characteristic features of these flows.

1 Introduction

An effective visual analysis of the dynamic behavior of 3D time-dependent flow fields is still a challenging problem in scientific visualization. Although a number of promising approaches have been introduced in recent years, the size and complexity of the data sets as well as the dimensionality of the underlying space-time domain makes the data handling, the analysis and the visual representation challenging and partially unsolved. In particular, it also proves to be inherently difficult to actually comprehend (in detail) the important characteristics of 3D time-dependent flow data.

In addition to others (streak lines, time lines, etc.), there exist two important kinds of characteristic curves for time-dependent flow fields: stream lines and path lines. While stream lines describe the steady behavior of the flow at a certain time step, path lines describe the paths of massless particles over time in the flow. Hence, the analysis of the dynamic behavior of flow fields is strongly related to the analysis of the behavior of the path lines.

One common approach to analyzing flow fields is to partition the flow domain into areas of characteristically different flow properties. To do so, a variety of different features have been proposed, such as topological features, vortical structures, or shock waves. They reflect different properties of the flow and therefore focus on the representation of different inherent structures. In fact, not all features may give useful information for every flow data set, and the selection of the relevant features is often left to the user in an unsupported way. Moreover, among the features there may be correlations which are either general due to their definition, or they occur in certain areas of particular flows and give relevant information about the behavior of the flow. Therefore we believe that not only the introduction and visualization of new features leads to a deeper understanding of the dynamic behavior of the flow field, but also an effective analysis of the relations between the features and the applications of these results for a visual representation. Our paper is one step along the recently challenging path towards a better understanding of 3D time dependent flow fields.

Our approach starts with the extraction of a number of properties (features, scalar values, and time series) at each point of a regular sampling of the 4D space-time domain. We have focused on properties describing the (local or global) behavior of the path lines, being either classical and well-established values in vector algebra, or properties newly proposed in this paper. The result of this step is a *path line attribute data set*: a four-dimensional multivariate data set collecting all computed path line properties.

The visual analysis of multidimensional multivariate data is a well researched topic in information visualization. A variety of techniques has been developed to visualizing such data sets making inherent correlations visible. Because of this we attempt to use information visualization approaches to analyzing the path line attributes data set. The results of this analysis (i.e., selections of path lines with certain combinations of properties) are then used for a focus+context visualization of either the selected path lines or the interesting properties. This way the user is able to do a simultaneous exploration in the 4D space-time domain of the flow and in the abstract path line attribute space. We show that this can give new insight into characteristic substructures of the flow which leads to a better understanding of time-dependent flow fields.

The rest of the paper is organized as follows. Section 2 mentions related work in the visualization of 3D time-dependent flow fields. Section 3 presents the properties of path lines which we extract for the further analysis. Section 4 describes our information visualization approach and explains how to use it for a focus+context visualization of the flow data. Section 5 applies our approach to a number of data sets. Section 6 draws conclusions and mentions issues of future research.

2 Related Work

The idea to segment a flow domain into areas of certain flow properties has been used for 3D steady flow fields for a variety of features, such as topological features [7, 13, 14, 25, 31] or vortex regions [11, 18, 16]. [20] provides a general framework of this in the context of topological features. [17] gives an overview on flow visualization techniques focusing on feature extraction approaches.

The extension of these techniques to 3D time-dependent fields is usually done by observing the feature regions over time, see [23, 22, 10] for topological features and [2, 3, 1, 24] for vortex features. Although these approaches provide insight into the flow behavior at arbitrary time steps, the analysis of the dynamic behavior based on path lines make specialized approaches necessary. [29] visualizes a number of carefully selected path lines to get static representations of the dynamic flow. [27] considers a segmentation of the flow domain based on local properties of the path lines. [28, 30] apply texture based visualization approaches to capture path line characteristics.

The idea of connecting information visualization and scientific visualization approaches is considered to be one of the "hot topics" in visualization [12]. Salzbrunn et al. published an approach of streamline predicates for steady flow [20]. The work closest to ours is the SimVis approach [4, 6] which uses approaches of information visualization to analyzing various kinds of simulation data. The main difference to our approach is that SimVis works on multiple scalar data describing certain properties of the simulation. Contrary to this, our approach works on dynamic flow data, focusing on local and global properties of path lines, i.e. on a multi-variate properties data set, derived from a 3D unsteady flow field.

3 Path Line Attributes

Given a 3D time-dependent vector field $\mathbf{v}(\mathbf{x}, t)$, $\mathbf{x}$ describes the 3D domain and t is the temporal component. Stream lines and path lines are generally different classes of curves [26]. Stream lines are the tangent curves of $\mathbf{v}$ for a fixed time t, while path lines describe the paths of massless particles in $\mathbf{v}$ over time.

Given a point $(\mathbf{x}, t)$ in the space-time domain, the stream line starting at $(\mathbf{x}, t)$ can be written in a parametric form

$$\mathbf{s}_{\mathbf{x},t}(\tau) = \mathbf{x} + \int_0^\tau \mathbf{v}(\mathbf{s}_{\mathbf{x},t}(s), t)\, ds \qquad (1)$$

while the path line starting at $(\mathbf{x}, t)$ has the parametric form

$$\mathbf{p}_{\mathbf{x},t}(\tau) = \mathbf{x} + \int_0^\tau \mathbf{v}(\mathbf{p}_{\mathbf{x},t}(s), s+t)\, ds. \qquad (2)$$

Path lines depict the trajectory of massless particles in a time-dependent flow. To characterize path lines, we consider two kinds of information: scalar values that describes local or global properties of a path line, and time series that collects information along a path line.

For scalar attributes, we compute a number of scalar properties of the path line starting at a given point $(\mathbf{x}, t)$ which reflect either local or global properties of the path lines. In the latter case, the value depends on the considered integration time. Since we are interested in the global behavior of the path lines, the integration time can be chosen rather large (relative to the time interval in which $\mathbf{v}$ is defined). In particular, we compute the scalar values in Table 1.

For time series we have investigated the attributes in Table 2.

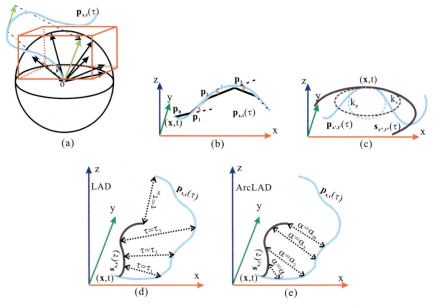

Fig. 1. a) Mapping the direction vectors along a path line to a unit sphere and calculating the bounding box approximation of the opening cone; b) Winding angle along a path line; c) Curvature difference between the path line and stream line pass through a specified point. d) LAD that records the Euclidean distance between the point of a path line the corresponding stream line at the same time τ; e) ArcLAD that records the Euclidean distance between the point of a path line and the corresponding stream line at the same arc length α from the start point

Table 1. Scalar Attributes

Id	Name	Description
$nonStraightV$	Non Straight Velocity	$\frac{\int_0^\tau \|\mathbf{v}(\mathbf{P}_{\mathbf{x},t}(s), s+t)\| \, ds - |\mathbf{P}_{\mathbf{x},t}(\tau) - \mathbf{x}|}{\tau}$
$distSE$	Relative start end distance	$\frac{|\mathbf{P}_{\mathbf{x},t}(\tau) - \mathbf{x}|}{\tau}$
$avDir$	Average direction	$\frac{\mathbf{P}_{\mathbf{x},t}(\tau) - \mathbf{x}}{\|\mathbf{P}_{\mathbf{x},t}(\tau) - \mathbf{x}\|}$
$avParticleV$	Average particle velocity	$\frac{\int_0^\tau \|\mathbf{v}(\mathbf{P}_{\mathbf{x},t}(s), s+t)\| \, ds}{\tau}$
$lyapunov$	Lyapunov exponent	$\frac{\log(\sqrt{\lambda_{max}(A^T A)})}{\tau}, A = \nabla_{\mathbf{x}} \mathbf{P}_{\mathbf{x},t}(\tau)$ [9, 19]
$wind$	Winding Angle	$\sum_{i=0}^{n-2} \angle((\mathbf{p}_{i+1} - \mathbf{p}_i), (\mathbf{p}_{i+2} - \mathbf{p}_{i+1}))$, Fig. 1b
lad	Local acceleration displacement	$\|\mathbf{P}_{\mathbf{x},t}(\tau) - \mathbf{s}_{\mathbf{x},t}(\tau)\|$, Fig. 1d
$curvDiff$	Curvature difference	$(\kappa_{\mathbf{s}} - \kappa_{\mathbf{p}})^2, \kappa_{\mathbf{p}} = \frac{\|\dot{\mathbf{p}} \times \ddot{\mathbf{p}}\|}{\|\dot{\mathbf{p}}\|^3}, \kappa_{\mathbf{s}} = \frac{\|\dot{\mathbf{s}} \times \ddot{\mathbf{s}}\|}{\|\dot{\mathbf{s}}\|^3}$, Fig 1c
div	Local divergence	$\mathrm{div}(\mathbf{v})$

Table 2. Time Series Attributes

Id	Name	Description
$DistEu$	Euclidean distance to start	$DistEu(\tau) = \|\mathbf{P}_{\mathbf{x},t}(\tau) - \mathbf{x}\|$
LAD	Local acceleration displacement	$LAD(\tau) = \|\mathbf{P}_{\mathbf{x},t}(\tau) - \mathbf{s}_{\mathbf{x},t}(\tau)\|$, Fig. 1d
$ArcLAD$	Arc local acceleration displacement	Fig. 1e
Dir	Direction vector	
$OpeningCone$	Opening cone	Fig. 1a
$Curvature$	Curvature	
$Velocity$	Velocity	

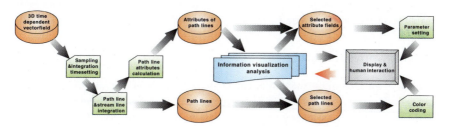

Fig. 2. Pipeline for analyzing path line attributes

4 System overview

Fig. 2 shows the pipeline of our path line attribute analysis approach. We start with a 3D time-dependent flow field **v** to be analyzed. As a first step, we apply a *sampling* of the space-time domain to obtain the points for which we compute the path line attributes. Note that since the data lives in a 4D domain, even a rather small sampling density may give a high amount of sample points. Therefore, the sampling density should be a compromise between the spatio-temporal accuracy of the analysis and the available computing resources. If the analysis delivers interesting features in certain smaller regions of the domain, this region can be analyzed using a higher sampling density to make sure the sampling rate is above the Nyquist frequency. At this state of the approach we also have to set the *integration time* for the path lines. Also this setting is a tradeoff between the fact that we want to have the path lines to be analyzed as long as possible and the property that most of the path lines should be integrated over the same time without leaving the domain.

The next step of the approach is the integration of the stream lines and path lines starting from the sampled points over the set integration time. For our examples we have used a 4th order Runge-Kutta integration. From these integrations we compute all path line attributes introduced in section 3.

The set of all path line attributes is the input of our information visualization core module which will be described in section 4.1 in more detail. Interactive visual analysis on the basis of state-of-the-art information visualization techniques and brushing in linked views is used to extract relevant correlations, interesting feature combinations, or general properties of the data. Note that the brushed features are not necessary physical variable. The result of this analysis is used to steering the visualization of the path lines and their attributes. If the interactive visual analysis delivers interesting features in a certain scalar path line attribute, we can visualize it using standard volume rendering techniques like direct volume rendering or slicing. Furthermore, the interactive visual analysis delivers a selection of interesting path lines having a certain combination of properties. They are visualized as 3D line structures with a color coded time component.

Our implementations of the visualization of the selected path lines and the selected attributes are based on Amira [21], whereas our information visualization analysis is based on the ComVis system which is described in section 4.1.

4.1 The ComVis system

ComVis is an interactive visualization tool. It supports conventional information visualization views such as 2D and 3D scatter plots, parallel coordinates, histograms, as well as a special curves view which is used for displaying function graphs. This combination of views makes it possible to analyze a wide variety of data where in the same row of a multi-variate table some values are scalar (just as it is usual) and others correspond to a function graph (common in various kinds of scientific data)[15]. The tool offers multiple linked views parallel to each other. Each view can be of any of the above mentioned view type. ComVis pays great attention to interaction. Due to advanced brushing and linking proved to be very powerful analytical tool. Users can brush the visualized data in any view, all linked views reflect the data selections by appropriate focus+context visualization. Furthermore, the user can use a simple, yet powerful line brush in the curves view. The line brush selects all curves which intersect the line. All brushes can be scaled and moved interactively. The multiple brush mode makes it possible to flexibly combine various brushes. The user selects brushes and boolean operations between them. AND, OR, and SUB are supported. Furthermore, the tool creates a composite brush in an iterative manner. This means that the user selects a current operation (AND, OR, or SUB) and draws a brush. The previous brushing state is combined with the new brush accordingly. The new state is computed, and it is used when the user draws another brush. In this way the user immediately gets visual feedback, and can very easily broaden the selection (using OR), or can further restrict the selection (using AND or SUB). Once the user is satisfied with a selection (or in the meantime), a tabular representation of the selected data can be shown and exported to file on demand.

5 Applications

We applied our approach to a number of data sets. Not surprisingly, not all attributes are interesting in all data sets, and different path line attributes turn out to be important for different data set. However, we can also identify several interesting coherencies between different path line attributes which seem to hold even for different data sets. Accordingly, we are optimistic that the here described analysis indeed provides a useful basis for future generalization of this approach.

5.1 3D time-dependent cylinder flow

Figures 3 and 4 present some results of analyzing a 3D time-dependent flow behind a circular cylinder. The cylinder is put in the origin with radius 0.5 and height 8.0, while the data set domain D is $[3.15, 19.74] \times [-2.06, 2.06] \times [0.09, 1.89] \times [0, 2\pi]$. This data set was kindly provided by Gerd Mutschke (FZ Rossendorf) and Bernd R. Noack (TU Berlin). We considered path lines at a $28 \times 14 \times 7 \times 6$ (191MB attribute file to ComVis) sampling and used an integration time of 1.5π (for the data set given in a 2π time slab). Figure 3a shows the direct volume rendering of one of the attribute fields *lyapunov*. In figure 3b, all path lines integrated from the sampled points are displayed. As we can see from figure 3a, there are certain patterns in the *lyapunov* attribute field. Low *lyapunov* values indicate stability of the path line. We use the information visualization approach to select the area with low *lyapunov*, as shown in the upper left of figure 4a. The visualization of the selected path lines is shown in figure 4b. Fig. 4c shows the seed area of the selected path lines at the time 0.

When investigating the visualized result, we can see that there are further different patterns in the low *lyapunov* path lines. It is obvious when we

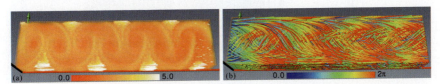

Fig. 3. Flow behind a cylinder: a) Direct volume rendering of the *lyapunov* attribute field at time 0; b) All considered path lines

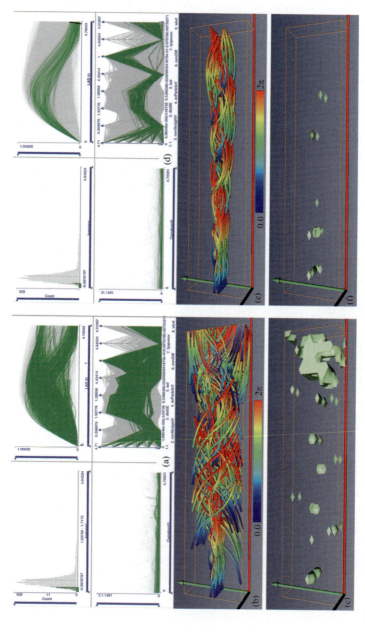

Fig. 4. Feature low *lyapunov* and *LAD*: a) Selecting low *lyapunov* area in ComVis; b) Visualization of selected path lines with low *lyapunov*; c) Visualization of the seeding area of the selected low *lyapunov* path lines at time 0; d) Selecting low *lyapunov* and parabola *LAD* area in ComVis; e) Visualization of selected path lines with low *lyapunov* and low *LAD*; f) Visualization of the seeding area of the selected low *lyapunov* and low *LAD* path lines at time 0

investigate the ComVis result of time series LAD, after choosing the cluster as shown in the upper right of figure 4d. We get the path line cluster whose LAD time series have small values at the end of the integration time. Fig. 4e and 4f present the visualization of the selected path lines and their seed areas. We notice that they stay in the middle of the domain and along the flow direction directly behind the cylinder.

5.2 Hurricane Isabel

Fig. 5 shows a visual analysis of the hurricane Isabel data set, which has been previously analyzed in a number of papers [8, 5]. We sample the domain with path lines at a resolution of $24 \times 24 \times 6 \times 6$ (253MB attribute file to ComVis), and set the maximum integration time to 30 hours (the whole data set covers 48 hours). Fig. 5a shows the visualization of all considered path lines. Fig. 5b show a direct volume rendering of $nonStraightV$ at time 0 (the starting time of the simulation).

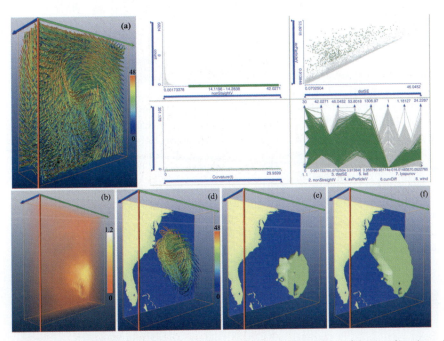

Fig. 5. Analysis and visualization of data set Hurricane Isabel: a) A visualization of all considered path lines. b) Direct volume rendering of the $nonStraightV$ attribute field at time 0; c) Selecting the area with high $nonStraightV$ which corresponds to swirling behavior in ComVis; d) Visualization of selected path lines of swirling behavior; e) Visualization of the seeding area of the selected swirling path lines at time 0; f) Visualization of the seeding area of the selected swirling path lines at all time steps

For this data set, we start the information visualization analysis, with the observation of the *avParticleV* vs. *distSE* scatter plot (upper right of Fig. 5c), showing a number of points on the diagonal but also a number scatter points clearly above it. We expect the points on the diagonal to represent path lines with a rather straight-line-like behavior, whereas the locations of the points above the diagonal may indicate a swirling behavior. Since *nonStraightV* is equivalent *avParticleV* vs. *distSE*, we selected all points above the diagonal, by considering points with a rather high *nonStraightV* (upper left of Fig. 5c). The parallel coordinate representation (lower right of Fig. 5c) shows that the selected path lines have a rather low *curvDiff*. This indicates that in these regions stream lines and path lines are locally rather similar. The curvature plot of the selected path lines doesn't have extreme values (lower left of Fig. 5c). The selected path lines are visualized in Fig. 5d, clearly showing that we have selected the ones swirling around the moving eye of the hurricane. Fig. 5e shows the areas where the selected path lines originate at time $t = 0$, while Fig. 5f shows the starting areas of the selected path lines for all time steps.

5.3 Airfoil

Figures 6 - 7 show a comparative visual analysis of 8 different data sets of a flow around an airfoil. The difference between these 8 data sets are the air injection frequency. The injection frequencies are 0(base), 0.2, 0.44, 0.6, 0.88, 1.0, 1.5 and 2.0. The goal of our analysis is to find the best air injection frequency which contributes the best lift power. It is known that abnormal vortex structures reduce the lift of the airfoil. Therefore, our visual analysis focuses on the areas with vortices where the probability of abnormal flow is high. We reduce our consideration to a small area around the areas with vortices.

We sample the interesting area with path lines at a resolution of $36 \times 12 \times 8 \times 10$ for each data set, and set the maximum integration time to 30 seconds (the whole data set covers different time domains for different frequencies and the path line integration will usually leave the domain within 30 seconds for each frequency). Fig. 6 shows the visualization of all considered path lines for

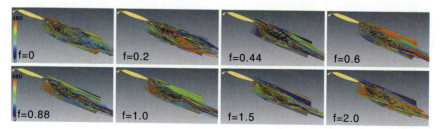

Fig. 6. The path lines started from the focus area of the airfoil flow field for different air injection frequency

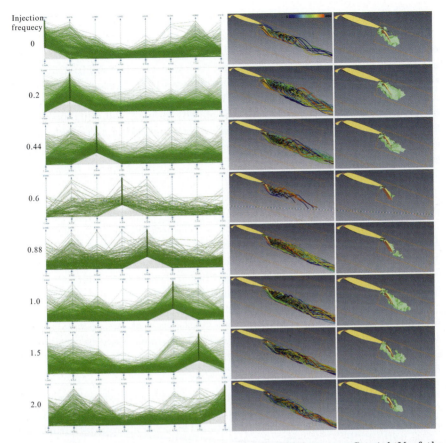

Fig. 7. The comparative analysis result for the attribute *nonStraightV* of the airfoil flow field for different air injection frequency. The pictures in the first column depict the selections of 70 percent highest *nonStraightV* for different frequencies in ComVis. The pictures in the second column depict the corresponding selected path lines for the first column. The pictures in the third column are the corresponding seeding area for the selections in the first column

different frequencies. We observe that most path lines behave well showing a rather straight behavior. The abnormal flows correspond to those non straight path lines. As our experience on these attributes, the *nonStraightV* is a good attribute to reflect the characteristics of straightness of path lines. So we compare this attribute computed at same location and same time for different frequency data sets in ComVis.

Fig. 7 shows the comparative result of the analysis of the *nonStraightV* for these 8 different frequencies. Relative analysis is popular in airfoil analysis since the relative flow behavior for different parts of an airfoil determines the lift power. We apply a relative selection here and select those path lines for each data set with 70 percent highest *nonStraightV* attributes. Those selected path lines and the corresponding seeding areas are visualized. We can see that these selected non straight path lines are closed to the area with vortices. And we can clearly observe that for a frequency 0.6, there are fewest non straight path lines and the non straight seeding areas are the smallest. So we find that for frequency 0.6, the probability of abnormal flow is less compared to others. We have tested several other percentage of the highest *nonStraightV*. All the results present the equivalent information. We conclude that 0.6 is the best air injection frequency among the 8 tests. The experience from the industry partner confirms this result successfully.

6 Conclusions

To getting insight into the dynamic behavior of path lines of 3D time-dependent flow fields is still a challenging problem for the visualization community. Path lines elude a straightforward extension from stream line based methods because path lines can be integrated only over a finite time, and they may intersect each other (at least when only considering their 3D reference locations). This paper is the - to the best of our knowledge - first approach to getting insight into the behavior of path lines by applying an approach from information visualization. In particular, we made the following contributions:

- We identified a number of local and global attributes of path lines which we expect to contain relevant information about the path line behavior.
- We interactively analyzed these attributes by using an approach from information visualization. The results were used to steering a 3D path line visualization.
- We applied our approach to a number of data sets, in order to get new insight into the path line behavior.

During our analysis it turned out that not all path line attributes gave useful results for all data sets. However, inherent and data independent correlations in the attribute data set can be expected, making a reduction of the attribute set possible. In particular, we have the impression that the investigation of path line attributes can indeed lead to a useful and practicable way of accessing/segmenting interesting flow features in time-dependent data sets, including swirling/vortical/rotating flow subsets, (e.g., via attributes *wind* and *nonStraightV*), quasi-steady flow structures, (e.g., via attributes *LAD* and *ArcLAD*, etc.), etc. We are optimistic with respect to these expectations, not at the least because it was, for example, fairly straight forward and quite easy to accomplish to extract the rotating main vortex of hurricane Isabel, which – to the best of our knowledge – cannot so easily be accomplished with any of the previously published vortex extraction methods.

Acknowledgments

We thank Bernd R. Noack for the supply of the cylinder data which was kindly provided by Gerd Mutschke. The Hurricane Isabel data produced by the Weather Research and Forecast (WRF) model, courtesy of NCAR and the U.S. National Science Foundation (NSF). Also many thanks to Jan Sahner for support and fruitful discussion. Parts of this work have been done in VRVis Research Center which is funded by an Austrian research program called Kplus. Also, parts of the work was funded by the Max-Planck center of visual computing and communication. All visualization of this paper have been created using ComVis (see http://www.vrvis.at/via/products/comvis.html) and Amira – a system for advanced visual data analysis [21] (see http://amira.zib.de/).

References

1. Bauer, D., Peikert, R.: Vortex tracking in scale space. In: Proc. VisSym 02, 233–240 (2002)
2. Banks, D.C., Singer, B.A.: Vortex tubes in turbolent flows: Identification, representation, reconstruction. In: Proc. IEEE Visualization 1994, 132–139 (1994)
3. Banks, D.C., Singer, B.A.: A predictor-corrector technique for visualizing unsteady flow. IEEE Transactions on Visualization and Computer Graphics., 1(2), 151–163 (1995)
4. Doleisch, H., Gasser, M., Hauser, H.: Interactive feature specification for focus+context visualization of complex simulation data. In: Proc. VisSym 03, 239–248 (2003)
5. Doleisch, H., Muigg, P., Hauser, H.: Interactive visual analysis of hurricane isabel. VRVis Technical Report (2004)
6. Doleisch, H., Mayer, M., Gasser, M., Priesching, P., Hauser, H.: Interactive feature specification for simulation data on time-varying grids. In: SimVis 05, 291–304 (2005)
7. Globus, A., Levit, C., Lasinski, T.: A tool for visualizing the topology of three-dimensional vector fields. In: Proc. IEEE Visualization 1991, 33–40 (1991)
8. Gruchalla, K., Marbach, J.: Immersive visualization of the hurricane isabel dataset. (2004)
9. Green, M.A., Rowley, C.W., Haller, G.: Detection of lagrangian coherent structures in 3D turbulence. Journal of Fluid Mechanics., 572, 111–120 (2007)
10. Garth, C., Tricoche, X., Scheuermann, G.: Tracking of vector field singularities in unstructured 3D time-dependent datasets. In: Proc. IEEE Visualization 2004, 329–336 (2004)
11. Hunt, J.C.R.: Vorticity and vortex dynamics in complex turbulent flows. Trans. Can. Soc. Mec. Engrs., 11:21 (1987)
12. Johnson, C.: Top scientific visualization research problems. IEEE Comput. Graph. Appl., 24(4), 13–17 (2004)
13. Löffelmann, H., Doleisch, H., Gröller, E.: Visualizing dynamical systems near critical points. In: Spring Conference on Computer Graphics and its Applications, 175–184 (1998)

14. Mahrous, K., Bennett, J., Scheuermann, G., Hamann, B., Joy, K.: Topological segmentation in three-dimensional vector fields. IEEE Transactions on Visualization and Computer Graphics., 10(2), 198–205 (2004)
15. Matkovic, K., Jelovic, M., Juric, J., Konyha, Z., Gracanin, D.: Interactive visual analysis end exploration of injection systems simulations. In: IEEE Visualization 2005, 391–398 (2005)
16. Peikert, R., Roth, M.: The parallel vectors operator - a vector field visualization primitive. In: Proc. IEEE Visualization 1999, 263–270 (1999)
17. Post, F.H., Vrolijk, B., Hauser, H., Laramee, R.S., Doleisch, H.: The state of the art in flow visualization: Feature extraction and tracking. Computer Graphics Forum., 22(4), 775–792 (2003)
18. Sujudi, D., Haimes, R.: Identification of swirling flow in 3D vector fields. AIAA Paper. 95,1715 (1995)
19. Shadden, S., Lekien, F., Marsden, J.E.: Definition and properties of lagrangian coherent structures from finite-time lyapunov exponents in two-dimensional aperiodic flows. Physica D., 212, 271–304 (2005)
20. Salzbrunn, T., Scheuermann, G.: Streamline predicates as flow topology generalization. In: Topo-In-Vis 2005 (2005)
21. Stalling, D., Westerhoff, M., Hege, H.C.: Amira: A highly interactive system for visual data analysis. The Visualization Handbook, 749–767 (2005)
22. Theisel, H., Seidel, H.P.: Feature flow fields. In: Proc. VisSym 03, 141–148 (2003)
23. Tricoche, X., Scheuermann, G., Hagen, H.: Topology-based visualization of time-dependent 2D vector fields. In: Proc. VisSym 01, 117–126 (2001)
24. Theisel, H., Sahner, J., Weinkauf, T., Hege, H.C., Seidel, H.P.: Extraction of parallel vector surfaces in 3d time-dependent fields and application to vortex core line tracking. In: Proc. IEEE Visualization 2005, 631–638 (2005)
25. Theisel, H., Weinkauf, T., Hege, H.C., Seidel, H.P.: Saddle connectors - an approach to visualizing the topological skeleton of complex 3D vector fields. In: Proc. IEEE Visualization 2003, 225–232 (2003)
26. Theisel, H., Weinkauf, T., Hege, H.C., Seidel, H.P.: Stream line and path line oriented topology for 2D time-dependent vector fields. In: Proc. IEEE Visualization 2004, 321–328 (2004)
27. Theisel, H., Weinkauf, T., Hege, H.C., Seidel, H.P.: Topological methods for 2D time-dependent vector fields based on stream lines and path lines. IEEE Transactions on Visualization and Computer Graphics., 11(4), 383–394 (2005)
28. Weiskopf, D., Erlebacher, G., Ertl, T.: A Texture-Based Framework for Spacetime-Coherent Visualization of Time-Dependent Vector Fields. In: Proc. IEEE Visualization 2003, 107–114 (2003)
29. Wiebel, A., Scheuermann, G.: Eyelet particle tracing - steady visualization of unsteady flow. In: Proc. IEEE Visualization 2005 (2005)
30. Weiskopf, D., Schramm, F., Erlebacher, G., Ertl, T.: Particle and Texture Based Spatiotemporal Visualization of Time-Dependent Vector Fields. In: Proc. IEEE Visualization 2005 (2005)
31. Weinkauf, T., Theisel, H., Hege, H.C., Seidel, H.P.: Boundary switch connectors for topological visualization of complex 3D vector fields. In: Proc. VisSym 04, 183–192 (2004)

Flow Structure based 3D Streamline Placement

Tobias Salzbrunn and Gerik Scheuermann

University of Leipzig {salzbrunn, scheuermann}@informatik.uni-leipzig.de

Summary. Visualizing vector fields using streamlines or some derived applications is still one of the most popular flow visualization methods in use today. Besides the known trade-off between sufficient coverage in the field and cluttering of streamlines, the typical user question is: Where should I start my streamlines to see all important behavior?

In previous work, we define flow structures as an extension of flow topology that permits a partition of the whole flow tailored to the users needs. Based on the skeletal representation of the topology of flow structures, we propose a 3D streamline placement generating a minimal set of streamlines, that on the one hand exactly illustrates the desired property of the flow and on the other hand takes the topology of the specific flow structure into account. We present a heuristic and a deterministic approach and discuss their advantages and disadvantages.

1 Introduction

Flow visualization is an important topic in scientific visualization. Various science and engineering disciplines apply many of its well-established methods. Especially streamlines are one of the most popular flow visualization methods in use. One of the key aspects determining the quality of such a visualization is to avoid occlusion of streamlines as much as possible while preserving as much of the desired information about the flow. Therefore a elaborate seeding strategy is of utmost importance. For 2D streamlines over a planar region and even for 2D manifolds in 3D there are two strategies that fulfill this goal satisfactorily. The image-guided strategy (e.g. Jobard et al. [4], Turk et al. [18]) tries to distribute streamlines evenly in space. The flow guided strategy (e.g. Verma et al. [19]) starts streamlines first at distinct features (here critical points) thus making them visible in the streamline pattern.

In 3D flow fields the clutter and occlusion problem is far more crucial. Mattausch et al. [8] extend the image guided approach to 3D flow using evenly spaced illuminated streamlines. They address the occlusion problem by mapping scalar flow features to streamline density. Additionally they work with depth cueing and halos along a streamline. Other works. e.g. Sadlo et al. [10] also use physical properties of the flow like regions of high vorticity. Westermann et al. [20] make use of level-set-methods relying on proximity of stream boundaries. Recently, Ye et al. [21] proposed a seeding strategy based on vector field topology with additional filtering via geometric and spatial properties. These contributions have in common, that they use flow features of interest to filter the streamlines and thus address the occlusion problem in contrast to the evenly-placement strategies used for 2D vector fields. But what in our view is lacking is a general framework that allows the user to clearly specify which behavior or feature of the flow should be the basis for the streamline seeding. A streamline seeding based on such a framework would result in a sparse representation of the flow tailored to the users needs.

In previous work [11], we define flow structures as an extension of flow topology. Flow structures allow to partition the whole flow according to specified properties. Based on these structures we propose a method for seeding streamlines that illustrates the desired property of the flow with a minimal set of streamlines while taking the topology of the specific flow structure into account. Our seeding strategy involves several steps. Based upon a given flow structure, we compute a skeletal representation of the topology of its parts. Based on this skeleton and various quality criteria, we calculate a minimal set of streamlines that exactly illustrates the desired property of the flow. Furthermore, we show that our approach can be adapted to gradually reduce the density of the streamlines in order to get an even sparser representation.

2 Related Work

Beside the previously discussed work on streamline placement, there is a wide variety of techniques for visualizing 3D flow. Post et al.[9] give a detailed overview of feature-extraction and tracking. Although this type of visualization barely uses streamlines but rather abstract icons as visualization primitives for local phenomena, the fundamental feature detection algorithms can be used to build flow structures as we showed in [12]. Another group of visualization uses dense and texture based techniques(Laramee et al. [6]). Although these techniques cover the whole flow, they are rather useful for 2D plane vector fields or 2D manifolds in 3D vector fields because of the inherent occlusion problem. There are also attempts to get a sparse representation of the flow by using clustering approaches (Heckel et al. [2] as well as Telea et al. [15]) or algebraic multigrids (Griebel et al. [1]). A different approach is flow topology (Helman et al. [3], Mahrous et al. [7], Theisel et al. [16], Tricoche et al. [17], and Scheuermann et al. [13]) that can also be used as basis for streamline seeding as shown by Ye et al. [21].

3 Construction of Flow Structures

As our streamline seeding strategy builds upon flow structures, we review the required formalism to construct such a flow structure. Let $D \subset \mathbf{R}^3$ be the domain of our steady three-dimensional flow. Its **velocity field** is a Lipschitz continuous map $v : D \rightarrow \mathbf{R}^3$, $x \mapsto v(x)$. A **streamline** of v passing through the point $x \in D$ is a map $S_x : J_x \rightarrow \mathbf{R}^3$ where $0 \in J_x \subset R$ is an interval of maximal extent and $S_x(0) = x$ and $\dot{S}_x(\tau) = v(S_x(\tau))$ $\forall \tau \in J_x$.

A **streamline predicate** SP is defined as a Boolean map on the set of all streamlines $\mathcal{S}$, i.e.

$$SP : \mathcal{S} \rightarrow \{ TRUE, FALSE \},$$
$$S \mapsto SP(S).$$

The corresponding **characteristic set** A_{SP} is defined as

$$A_{SP} = \bigcup_{S_x \in \mathcal{S},\ SP(S_x)=TRUE} S_x(J_x) \subset D$$

We need a grouping mechanism that creates a finite number of groups of streamlines with common properties. We call the result of this grouping a **flow structure**. Our mechanism assumes a finite set $\mathcal{G}$ of streamline predicates with disjunct characteristic sets, i.e.

$$\mathcal{G} = \{\ SP_\lambda \mid \lambda \in \Gamma\ \}, \quad A_{SP_\lambda} \cap A_{SP_\mu} = \emptyset \ \forall \lambda, \mu \in \Gamma,\ \lambda \neq \mu,$$

where Γ is an index set. As **flow structure** we define the partition

$$D = \bigcup A_{SP_\lambda}.$$

4 Flow Structure Examples

Our first dataset we use for the examples results from a simulation of the flow around a sphere with a drilled hole in the center. The hole substantially changes the flow behavior. The sphere center is located at the origin and it has a diameter of 200. The underlying unstructured grid contains 2.5 million tetrahedra. For discretization we sample a representative finite subset $\tilde{\mathcal{S}}$ of all streamlines $\mathcal{S}$. Therefor we use a Cartesian grid in the area $[-250; 250] \times [-125; 125] \times [-125; 125]$ with a spacing of 6.25 in all directions as starting positions. This is a set of 136,161 streamlines that fills the space around the ball in a dense manner.

The second dataset we use corresponds to a single time step of an unsteady simulation of the German train ICE. The train travels at a velocity of about 250 km/h with a wind blowing from the side at an angle of 30 degrees. The wind causes vortices to form on the lee side of the train, creating a drop in pressure that has adverse effects on the trains track holding. For our computations we choose a region of interest around the front wagon. To get a finite subset $\tilde{S}$ of S, we take a Cartesian grid in the area $[-15000, 45000] \times [-11000, 21000] \times [400, 5200]$ with 400 units as spacing in all directions as starting positions.

We use three flow structures as examples for our streamline seeding. First we want to study the deviation of the flow from the principal input flow direction. We obtain the deviation by integrating the difference between tangent vector direction and main inflow direction along the streamlines. Then we compute the deviation and take as minimum deviation $d_{min} = 0.1$. We define the streamline predicate: D - Deviation of $\tilde{S}$ from a given direction greater than d_{min}. The discretized characteristic set is comprised of all voxels visited by a streamline fulfilling the respective streamline predicate. For a formal description of this kind of discretization we refer the reader to [3]. The resulting simple flow structure is $\mathcal{G}_{Dev} = \{ A_D, \bar{A}_D \}$. Figure 1 shows the main flow direction and the characteristic set A_D for $d_{min} = 0.1$ (weak deviation) and $d_{min} = 0.6$ (strong deviation).

Second we want to examine the swirling behavior of flow with respect to its main vortices. For this purpose we use the streamline predicate developed in [12]. It states, whether a given streamline swirls around a specific vortex v. We apply the predicate on two different vortices resulting in two flow structures $\mathcal{G}_{v1} = \{ A_{v1}, \bar{A}_{v1} \}$ and $\mathcal{G}_{v1} = \{ A_{v2}, \bar{A}_{v2} \}$ (with A_{vi} being the characteristic set containing all streamlines that swirl around vortex i). The discretization is similarly to the first predicate. Figure 2 shows the vortex core lines of the flow using the Sujudi-Haimes method[14]. The dominant flow pattern are three ring

Fig. 1. Main flow direction and corresponding characteristic sets A_D for $d_{min} = 0.1$ (weak deviation) and $d_{min} = 0.6$ (strong deviation) (from left to right)

Fig. 2. Left image: Vortex cores lines of used dataset (according to the method of Sujudi-Haimes [14]) after a filtering step. There are three main ring vortices. Middle and right picture: Characteristic sets A_{v1} and A_{v2}

vortices on the lee side of the sphere. We label the inner ring vortex as vortex $v1$ and the middle ring vortex as vortex $v2$. The corresponding characteristic sets A_{v1} and A_{v2} are also shown in figure 2.

5 Streamline Seeding

In most cases the number of streamlines belonging to a characteristic set is considerable less than the number of all streamlines from $\tilde{S}$. However, showing all streamlines of a characteristic set still tends to produce visual clutter. Therefore, we need a strategy to choose a minimal representative set of streamlines from a characteristic set. Looking upon the respective discretized characteristic set as a region, we can get a representation that captures the essential topology of this region in an easy to understand and very compact form using a skeleton representation (medial surface). We compute the skeleton of the discretized characteristic set with a thinning approach as suggested in [5]. Figure 3 shows the discretized characteristic set of A_D (left image) and its corresponding skeleton (right image).

Every voxel of the resulting skeleton represents all the voxels of the discrete characteristic set that have no shorter distance to any of the other skeleton voxel. To get this set of voxels for every skeleton voxel we use a flood-fill-algorithm starting for the first run with the skeleton voxels themselves and we continue by assigning the next nearest neighbors (i.e. max 26 voxels) to the respective skeleton voxels for the next runs. In case of conflicting assignments concerning two skeleton voxels we have to calculate the actual distance and assign the voxel to the skeleton voxel with the shortest distance. If the distance is equal we take the first skeleton voxel. After several runs we get a partition of the discrete characteristic set according to the assignment to a skeleton voxel.

We define that a streamline **_visits_** a skeleton voxel if the streamline intersects the skeleton voxel itself or one of the voxels assigned to this skeleton voxel. A set of streamlines from a discrete characteristic set that visits all its skeleton voxels shows (approximately) its topological structure and hence can be seen as a representative set of streamlines. We call such a set of streamlines

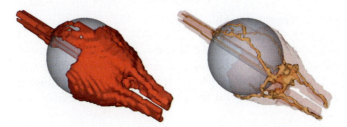

Fig. 3. Left side shows the boundary of the discrete characteristic set of A_D. The skeleton of this set together with the highly transparent boundary is shown on the right side

a *configuration*. Its obvious that the characteristic set itself is a configuration. What we are looking for is a minimal configuration, i.e. a configuration with as less streamlines as possible. Furthermore, every skeleton voxel should be visited only once in the ideal case. This is not possible in most cases, since most characteristic sets have sub-branches originating from a common main branch. Hence, every skeleton voxel representing this main branch is visited by all the streamlines coming from the separate sub-branches. That means, one can only try to minimize the number of multiple visits.

5.1 Heuristic approach

To compute a minimal configuration with minimal multiple visits, we use at first a heuristic approach. For the first run one of the skeleton voxels is randomly chosen. As a skeleton voxel is always a voxel of the discrete characteristic set, there is at least one streamline of the characteristic set that intersects this skeleton voxel. We take a fitting streamline and note which skeleton voxels are visited during its course. For every skeleton voxel a counter is installed which is incremented if the voxel is visited (a streamline can increment a counter only once). For the next run a skeleton voxel not visited by previous streamlines (i.e. the respective counter is zero) is used to get the next streamlines. This process is repeated as long as an unvisited skeleton voxel exists. In a last step we sum up all counters and subtract the number of skeleton voxel in order to get the number of multiple visits. This quality measure is saved together with the computed configuration. In this way a number of configurations is computed and the one with the best quality measure is used for the streamline placement.

On top of the requirement that every branch of the characteristic set is represented by a streamline, one could demand that a representative streamline additionally should not be at the border of a characteristic set but be instead in the center. Looking at the discrete case, this translates into the requirement that the streamlines of a configuration should be as near to the skeleton as possible. Hence, another quality criteria could be the summed up distances of the streamlines with respect to the skeleton voxels.

5.2 Deterministic approach

The heuristic approach does not guarantee to find the optimal solution. Its main advantage is, that the runtime depends beside the number of configurations mainly on the number of skeleton voxels. A deterministic algorithm has to systematically compute all possibilities to find the guaranteed best configuration:

Starting from the set of all unvisited skeleton voxels, for each streamline the number of visited skeleton voxels from this set is counted. The streamline with the largest number of visited skeleton voxels is moved from the set of all streamlines to the configuration. This has to be repeated as long as there are unvisited skeleton voxels.

The drawback of the deterministic algorithm is, that for each iteration its runtime depends on the number of streamlines. This number tends to be exceedingly large for characteristic sets with high resolution. Tables 1-4 compare the heuristic and the deterministic approach by means of computation time and quality of the resulting configuration of the characteristic sets discussed in section 7. While on the one hand the deterministic algorithm needs proportional more computation time, on the other hand the heuristic algorithm converges more slowly to the optimum for increasing numbers of streamlines. (Note that the heuristic algorithm is optimized with respect to multiple visits.)

6 Sparse Seeding

Even if there is a minimal configuration, the number of streamlines could still be too high to avoid cluttering. How many streamlines a minimal configuration has is basically determined by the number of skeleton voxels. This number is determined again by the shape of the characteristic set, but also by the resolution of the corresponding discrete characteristic set. Hence, one can gradually use a coarser resolution for the discrete characteristic set to get

Table 1. Streamline placement for characteristic set A_{v1} (368 streamlines, 3957 voxels, and 609 skeleton voxels): heuristic (100 configurations) and deterministic approach in comparison

	computation time	# streamlines	# multiple visits
Deterministic	1.86[s]	21	8798
Heuristic (best/worst)	4.65[s]	22(29)	8309(12235)

Table 2. Streamline placement for characteristic set A_{v2} (2719 streamlines, 7237 voxels, and 521 skeleton voxels): heuristic (100 configurations) and deterministic approach in comparison

	computation time	# streamlines	# multiple visits
Deterministic	33.67[s]	21	12278
Heuristic (best/worst)	35.01[s]	24(27)	13127(27354)

Table 3. Streamline placement for characteristic set $\bar{A}_D$ (131062 streamlines, 119207 voxels, and 7553 skeleton voxels): heuristic (100 configurations) and deterministic approach in comparison

	computation time	# streamlines	# multiple visits
Deterministic	37.51[s]	13	9883
Heuristic (best/worst)	39.38[s]	19(21)	5527(19859)

Table 4. Streamline placement for characteristic set A_D (4339 streamlines, 9000 voxels, and 328 skeleton voxels): heuristic (100/200 configurations) and deterministic approach in comparison

	computation time	# streamlines	# multiple visits
Deterministic	40.05[m]	347	27210
Heuristic 100 (best/worst)	18.55[m]	531(576)	45367(49775)
Heuristic 200 (best/worst)	35.75[m]	526(558)	44845(50416)

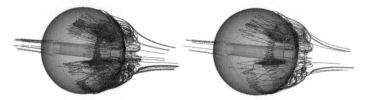

Fig. 4. Using different resolutions of voxelization. A coarser voxelization of characteristic sets yields to a sparser streamline placement. Both pictures show the streamlines of the minimal configuration of the characteristic set A_{v2}. The different voxel sizes (left 6.25, right 12.5) of the discrete characteristic set yields two different streamline densities

less streamlines. This reduction is not for free, since structures of the characteristic set from the size 2δ and up need a voxel size of δ. That means one has to find a compromise between the accuracy of the representation of the characteristic set of the streamlines and the number of streamlines used. Figure 4 shows a minimal configuration of the characteristic set A_{v2} with different voxel sizes (6.25 and 12.5) for the discrete characteristic set. The double voxel size yields to a significant reduction of streamlines while still representing the main topological structures of A_{v2}.

7 Results

We first apply our streamline seeding on the single characteristic set A_{v1}. The set is voxelized with a voxel size of 6.25. The pictures in the first row of figure 5 show all streamlines of A_{v1}. From this set, we have to choose a representative subset. As quality criteria we use the number of multiple visits. The pictures of the next two rows show the best (red streamlines) and the worst (blue streamlines) minimal configuration from 500 configurations. The two configurations only differ in detail. The worst configuration is more dense around the vortex cores, because each extra circle around the inner vortex core line probably increases the number of multiple visits. The pictures of the last row show a random selection of streamlines from A_{v1} using the same number of streamlines as in the best minimal configuration. The comparison of the placement shows that it is useful to invest computing time in a methodical selection of streamlines.

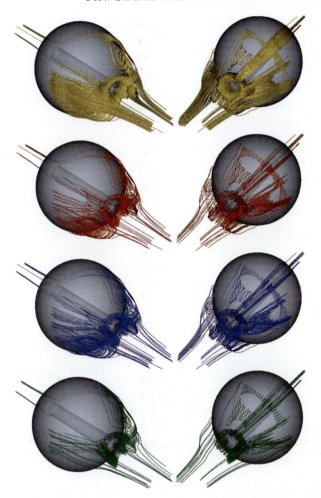

Fig. 5. Streamline seeding based on flow structure $\mathcal{G}_{v1}$. The first row shows all streamlines of characteristic set A_{v1} (i.e. streamlines circling around the inner ring vortex). The next two rows show the best (red) and worst (blue) streamline placement according to our quality criteria. For a comparison the last row shows a streamline placement by random selection of streamlines from A_{v1} (number of streamlines equal to that of the best placement) The two columns show different perspectives

In a second example we construct a streamline placement for the whole flow. But instead of a regular seeding of the flow volume, we want to amplify the deviation of the flow from the main direction. Therefore, we partition the flow with the flow structure $\mathcal{G}_{Dev} = \{\ A_D, \bar{A}_D\ \}$ introduced in section 3. The picture of the first three rows of figure 6 show the best minimal configuration for the characteristic sets A_D (red streamlines), $\bar{A}_D$ (purple streamlines),

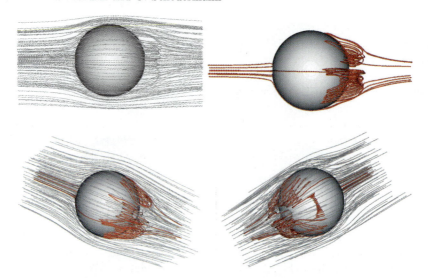

Fig. 6. Streamline placement for the whole flow using a partition by flow structure $\mathcal{G}_{Dev}$. The discrete characteristic sets are voxelized with voxel size 12.5. The first row shows the minimal configuration of the characteristic set A_D (left) and $\bar{A}_D$ (right). The second row shows both resulting sets of streamlines combined using two other viewpoints

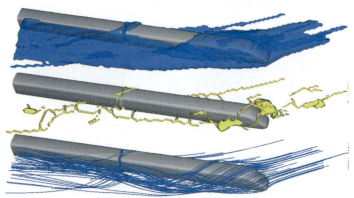

Fig. 7. Streamline placement based on flow structure $\mathcal{G}_{Dev}$ applied to the ICE-dataset (top down): characteristic set A_D, skeleton, and streamline placement

and both combined. Because the streamlines are still very dense, we use additionally a coarser voxelization of the characteristic sets. The resulting improvements concerning streamline density can be seen in direct comparison on the picture of the right column. The pictures of the last row show the best minimal set with coarse voxelization from to different perspectives. Although the occlusion of the streamlines is very modest, one can approximately see the parts of the flow where the deviation from the main flow is high. This results from the tailored flow structure the streamline placement was based on. Of course many other streamline placements of the flow with a different emphasis of a flow behavior are possible. The user has just to define and compute a fitting flow structure as basis for a new streamline placement - a streamline placement tailored to the users needs.

In the last example we apply the flow structure $\mathcal{G}_{Dev}$ to the ICE-dataset. Figure 7 shows the corresponding characteristic set A_D, the skeleton, and the streamline placement. The resulting streamlines (93 from a total of 4045) represent the structure of the characteristic set in a good fashion.

8 Conclusion

We introduced a new streamline seeding method based on flow structures. Applied to two realistic CFD-datasets, the method proved to bring out a minimal set of streamlines that illustrate the desired flow behavior with modest occlusion. We compared a heuristic and a deterministic approach. Depending on the properties of the respective characteristic set (number of skeleton voxels, number of streamlines) both approaches make sense. We also showed how to reduce the streamline density step by step in order to achieve an even sparser representation.

Acknowledgment

This work was partly supported by DFG grant SCHE 663/3-7. We also thank the reviewers for their valuable suggestions.

References

1. M. Griebel, T. Preusser, M. Rumpf, M.A. Schweitzer, and A. Telea. Flow Field Clustering via Algebraic Multigrid. In *IEEE Visualization 2004*, pages 35–42, Austin, Texas, 2004.
2. B Heckel, G.H. Weber, B. Hamann, and K.I. Joy. Construction of Vector Field Hierarchies. In *IEEE Visualization 1999*, pages 19–25, San Francisco, CA, 1999.
3. J. L. Helman and L. Hesselink. Visualizing Vector Field Topology in Fluid Flows. *IEEE Computer Graphics and Applications*, 11(3):36–46, May 1991.
4. B. Jobard and W. Lefer. Creating evenly-spaced streamlines of arbitrary density. In *Visualization in Scientific Computing '97 (Proceedings of the 8th Eurographics Workshop on Visualization in Scientific Computing '97*, pages 43–55, 1997.
5. A. Kuba and K. Palagyi. A 3D 6-Subiteration Thinning Algorithm for Extracting Medial Lines. *Pattern Recognition Letters*, 19(7):613–627, 1998.
6. R.S. Laramee, H. Hauser, H. Doleisch, B. Vrolijk, F.H. Post, and Weiskopf D. The State of the Art in Flow Visualization: Dense and Texture Based Techniques. *Computer Graphics Forum*, 23(2):203–221, 2004.

7. K. Mahrous, J. Bennett, G. Scheuermann, B. Hamann, and K. I. Joy. Topological Segmentation of Three-Dimensional Vector Fields. *IEEE Transactions on Visualization and Computer Graphics*, 10(2):198–205, 2004.
8. O. Mattausch, T. Theußl, H. Hauser, and E Gröller. Strategies for Interactive Exploration of 3d Flow Using Evenly-Spaced Illuminated Streamlines. In *Proceedings of Spring Conference on Computer Graphics*, pages 213–222, 2003.
9. F.H. Post, B. Vrolijk, H. Hauser, R.S. Laramee, and H. Doleisch. The State of the Art in Flow Visualization: Feature Extraction and Tracking. 22(4):775–792, 2003.
10. F. Sadlo, R. Peikert, and E. Parkinson. Vorticity Based Flow Analysis and Visualization for Pelton Turbine Design Optimization. In *IEEE Visualization 2004*, pages 179–186, Austin, Texas, 2004.
11. T. Salzbrunn and G. Scheuermann. Streamline Predicates As Flow Topology Generalization. In *Topo-In-Vis Proceedings 2005*, 2005.
12. T. Salzbrunn and G. Scheuermann. Streamline Predicates. *IEEE Transactions on Visualization and Computer Graphics*, 12(6):1601–1612, 2006.
13. G. Scheuermann, K.I. Joy, and W. Kollmann. Visualizing Local Vector Field Topology. *Journal of Electronic Imaging*, 9:356–367, 2000.
14. D. Sujudi and R. Haimes. Identification of Swirling Flow in 3D Vector Fields. Technical Report AIAA Paper 95–1715, American Institute of Aeronautics and Astronautics, 1995.
15. A. Telea and J.J. van Wijk. Simplified representation of vector fields. In *IEEE Visualization 1999*, pages 35–42, San Francisco, CA, 1999.
16. H. Theisel, T. Weinkauf, H.C. Hege, and H.P. Seidel. Saddle Connectors - An Approach to Visualizing the Topological Skeleton of Complex 3d Vector Fields. In *IEEE Visualization 2003*, pages 225–232, 2003.
17. X. Tricoche, T. Wischgoll, G. Scheuermann, and H. Hagen. Topological Tracking for the Visualization of Timedependent Two-Dimensional Flows. *Computers & Graphics*, 26(2):249–257, 2002.
18. G. Turk and D. Banks. Image-Guided Streamline Placement. In *Computer Graphics Annual Conference Series*, pages 453–460, 1996.
19. V. Verma, D. Kao, and A. Pang. Flow-Guided Streamline Seeding Strategy. In *IEEE Visualization 2000*, pages 163–170, Salt Lake City, Utah, 2000.
20. R. Westermann, C. Johnson, and T. Ertl. A Level-Set Method for Flow Visualization. In *IEEE Visualization 2000*, pages 147–154, Salt Lake City, Utah, 2000.
21. X. Ye, D. Kao, and A. Pang. Strategy For Seeding 3d Streamlines. In *IEEE Visualization 2005*, pages 471–478, Minneapolis, MN, 2005.

Critical Points of the Electric Field from a Collection of Point Charges

Nelson Max[1] and Tino Weinkauf[2]

[1] Lawrence Livermore National Laboratory, 7000 East Avenue, Livermore, CA, 94550, USA, `max2@llnl.gov`
[2] Zuse Institute Berlin, Takustr. 7, D-14195 Berlin, Germany, `weinkauf@zib.de`

Summary. The electric field around a molecule is generated by the charge distribution of its constituents: positively charged atomic nuclei, which are well approximated by point charges, and negatively charged electrons, whose probability density distribution can be computed from quantum mechanics (Atoms in Molecules: A Quantum Theory, Clarendon, Oxford, 1990). For the purposes of molecular mechanics or dynamics, the charge distribution is often approximated by a collection of point charges, with either a single partial charge at each atomic nucleus position, representing both the nucleus and the electrons near it, or as several different point charges per atom.

1 Introduction

The critical points in the electric field are useful in visualizing its geometrical and topological structure, and can help in understanding the forces and motion it induces on a charged ion or neutral dipole. Most visualization tools for vector fields use only samples of the field on the vertices of a regular grid, and some sort of interpolation, for example, trilinear, on the grid cells. There is less risk of missing or misinterpreting topological features if they can be derived directly from the analytic formula for the field, rather than from its samples. This work presents a method which is guaranteed to find all the critical points of the electric field from a finite set of point charges. To visualize the field topology, we have modified the saddle connector method of [Th03] to use the analytic formula for the field.

The analysis below would also apply to gravity in astronomy, since the Newtonian gravitational force outside a spherically symmetric body is the same as if all its mass were concentrated at its center.

2 Electric Potential and Field

The electrostatic potential $P(X)$ at position X, resulting from a point charge q at position A is

$$P(X) = \frac{q}{4\pi\epsilon|X-A|} = \frac{q}{4\pi\epsilon((X-A)\cdot(X-A))^{1/2}}$$

where ϵ is the permittivity of the medium. Since we are only interested in the field topology here, we will for simplicity below neglect the factor $4\pi\epsilon$. By definition, the electrostatic field F is the negative gradient of the potential, so, neglecting the factor $4\pi\epsilon$,

$$\begin{aligned} F(X) &= -\nabla \frac{q}{((X-A)\cdot(X-A))^{1/2}} \\ &= \frac{q(X-A)}{((X-A)\cdot(X-A))^{3/2}} \cdot \end{aligned} \quad (1)$$

Therefore, for a set of n charges q_i at positions A_i, the electrostatic field is

$$F(X) = \sum_{i=1}^{n} \frac{q_i(X-A_i)}{((X-A_i)\cdot(X-A_i))^{3/2}} \cdot \quad (2)$$

3 Octree Method for Finding Critical Points

Our method of finding critical points of F starts with a single cubical cell, or a regular grid of such cells, guaranteed as in section 4 to contain all the critical points of F. We describe below a test to prove that a cubical cell C contains no critical points of F. This test is based on computing component-wise bounds F^{max} and F^{min} for the possible values of $F(X)$ within the cell. In the column vectors F^{max} and F^{min}, if any of the three components are both positive or both negative, we know that $F(X)$ cannot be zero for any X in C, so there are no critical points of the field in C, the test passes, and the cell can be skipped. Otherwise, all of the components of $F(X)$ could possibly become zero inside C, and the test fails, so we divide the cell into eight equal subcells, and recursively test these subcells. Thus we converge on the critical points. At the limiting depth of the recursion, the centers of cells which fail the test are written to an output list. This list is then pruned by averaging clusters of points within a threshold distance of each other. (Currently, this threshold is taken as 6 times the edge length of the cell at the recursion limit.) The test will fail for cells containing any of the point charge locations, so these locations will also be on the output list. The octree is never explicitly stored, and the only memory it uses is for the portion in the stack for the recursion. By trading space for time, one could store the force at vertices in some local portion of the octree, saving some recomputation for neighboring cells.

Consider a cubical cell C of side s, with vertices V_{ijk}, where i, j, and k are either 0 or 1, and count along the x, y, and z axes respectively. Our goal is to get bounds F^{max} and F^{min} on the values of $F(X)$ inside this cell, using the values of F on the eight vertices V_{ijk}, and bounds on the derivatives of F inside C. Since F is the negative gradient of the potential P, its derivatives are the entries in H, the negative of the Hessian matrix of second derivatives of the potential, *i.e.*

$$\mathsf{H}(X) = \sum_{i=1}^{n} \left(\frac{q_i \mathsf{I}}{|X - A_i|^{3/2}} - 3 \frac{q_i (X - A_i)(X - A_i)^\top}{|X - A_i|^{5/2}} \right). \tag{3}$$

Here I is the identity matrix, and $(X - A_i)(X - A_i)^\top$ is a square matrix, since X and A_i are column vectors. Let H^j be the jth row of H and F^j be the jth component of the vector F. Let V be one of the vertices V_{ijk}, and let X be any point inside the cell. Applying the mean value theorem to the function $E^j(t) = F^j(V + t(X - V))$, for $j = 1, 2$, or 3, there is a value t_j with $0 \le t_j \le 1$, so that

$$F^j(X) = F^j(V) + \mathsf{H}^j(Y_j)(X - V) \tag{4}$$

where $Y_j = V + t_j(X - V)$. Therefore bounds on the entries of H will help us get bounds on the components of F.

Consider the denominators in equation (3). The quantity $|Y - A_i|$ attains its maximum in C at one of the vertices of the cube C. If A_i is not inside cell C, then $|Y - A_i|$ attains a positive minimum in C at a vertex of C, or at the projection of A_i on one of the faces or edges of C. (Cells containing a point charge A_i fail the test, and are recursively subdivided.)

For each point charge A_i and each of the 9 elements of H, we can find the upper and lower bounds for each of the two terms in equation (3), using either the minimum or the maximum value of the denominator, taking into account the multiple possible signs of the numerator, and its upper and lower bounds. By summing the bounds on these two terms over all the point charges, we can get element-wise upper and lower bounds H^{max} and H^{min} for the matrix $\mathsf{H}(Y)$.

Then, using each vertex V_{ijk} in equation (4), we can obtain component-wise upper and lower bounds F_{ijk}^{max} and F_{ijk}^{min} for $F(X)$ with X in C, by taking into account the signs of the components of the column vector factor $(X - V)$ in equation (4). For example, if V in equation (4) is V_{011}, then for X in C, $X - V$ is of the form $(x, y, z)^\top$, with $0 \le x \le s$, $-s \le y \le 0$, and $-s \le z \le 0$, so, as shown in 1D in figure 1, we get the component-wise bounds

$$F_{011}^{max} = F(V_{011}) + \max(0, \mathsf{H}^{max}(s, 0, 0)^\top) + \max(0, \mathsf{H}^{min}(0, -s, 0)^\top)$$
$$+ \max(0, \mathsf{H}^{min}(0, 0, -s)^\top),$$

and

$$F_{011}^{min} = F(V_{011}) + \min(0, \mathsf{H}^{min}(s, 0, 0)^\top) + \min(0, \mathsf{H}^{max}(0, -s, 0)^\top)$$
$$+ \min(0, \mathsf{H}^{max}(0, 0, -s)^\top).$$

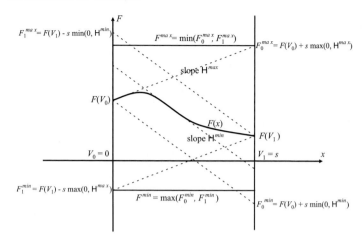

Fig. 1. A 1D example for computing bounds on a function $F(x)$

Finally, let $F^{max} = \min_{ijk} F^{max}_{ijk}$ and $F^{min} = \max_{ijk} F^{min}_{ijk}$ be the componentwise minimum and maximum from these eight respective estimates, giving the tightest bounds they collectively impose on $F(X)$ in C. As described above, these bounds are used to limit the octree search. Note that in the 1D case illustrated in figure 1, $F^{max} > 0$, and $F^{min} < 0$, so further subdivision would be required. However, the lowest vertex of the parallelogram bounded by the dotted lines of slope H^{max} and H^{min} from $F(V_0)$ and $F(V_1)$ is above the $F = 0$ axis, so we could actually skip this cell. In 3D, the range of F values satisfying the eight pairs of linear inequalities could be found by low dimensional linear programming [deB00], but this is more complicated, and so far we have not done so.

4 Finding a Sphere Containing all the Critical Points

In order to make sure that all the critical points are found, we need to make the initial root cube or grid large enough to contain them all. We instead find a large enough sphere, and enclose it in the initial grid. To do this, we expand equation (2) into a series of terms, which decrease with different inverse powers of the distance $r = |X|$ of the point X from the origin. This series is closely related to the multipole expansion of the potential described in [Wiki], [Ja62], or [Sch94], but is slightly different because we are expanding the field, and not the potential. We will find a sphere outside of which the first non-vanishing term of this series dominates the sum of the later terms, so that $F(X)$ cannot vanish.

Consider first a single point charge q at position A, as in equation (1), let $r = |X|$, and rewrite the denominator as follows:

$$\begin{aligned}
F(X) &= \frac{q(X-A)}{((X-A)\cdot(X-A))^{3/2}} \\
&= \frac{q(X-A)}{(|X|^2 - 2A\cdot X + |A|^2)^{3/2}} \\
&= \frac{q(X-A)}{(r^2(1+\frac{1}{r^2}(-2A\cdot X + |A|^2))^{3/2}} \\
&= \frac{qr(\frac{X}{r} - \frac{A}{r})}{r^3\left(1 + \left(\frac{|A|^2}{r^2} - 2\frac{A}{r}\cdot\frac{X}{r}\right)\right)^{3/2}} \\
&= \frac{q}{r^2}\left(\hat{X} - \frac{A}{r}\right)\left(1 + \left(\frac{|A|^2}{r^2} - 2\frac{A}{r}\cdot\hat{X}\right)\right)^{-3/2}, \quad (5)
\end{aligned}$$

where $\hat{X} = X/r$ is the unit vector in the direction of X. Let

$$t = \frac{|A|^2}{r^2} - 2\frac{A}{r}\cdot\hat{X}.$$

Then, expanding $(1+t)^{-3/2}$ by the binomial series (the same as its Taylor expansion about $t = 0$), which converges for $|t| < 1$, we get

$$\begin{aligned}
(1+t)^{-3/2} &= \left(1 + (-3/2)t + (-3/2)(-5/2)t^2/2! + \cdots\right) \\
&= \sum_{l=0}^{\infty}\binom{-\frac{3}{2}}{l}t^l \quad (6)
\end{aligned}$$

where the generalized binomial coefficient

$$\binom{s}{l} = \frac{s(s-1)(s-2)\cdots(s-l+1)}{l!}.$$

Substituting this binomial series into equation (5), and writing $A\cdot\hat{X}$ as $A^\top\hat{X}$, we have

$$\begin{aligned}
F(X) &= \frac{q}{r^2}\left(\hat{X} - \frac{A}{r}\right)\left(1 + \left(-\frac{3}{2}\right)t + \left(-\frac{3}{2}\right)\left(-\frac{5}{2}\right)\frac{t^2}{2!}\right. \\
&\quad \left. + \cdots + \left(-\frac{3}{2}\right)\left(-\frac{5}{2}\right)\cdots\left(-\frac{2l+1}{2}\right)\frac{t^l}{l!} + \cdots\right) \quad (7) \\
&= \frac{q}{r^2}\left(\hat{X} - \frac{A}{r}\right)\left(1 + \left(-\frac{3}{2}\right)\left(\frac{|A|^2}{r^2} - 2\frac{A^\top\hat{X}}{r}\right)\right. \\
&\quad \left. + \left(-\frac{3}{2}\right)\left(-\frac{5}{2}\right)\left(\frac{|A|^2}{r^2} - 2\frac{A^\top\hat{X}}{r}\right)^2/2 + \cdots\right)
\end{aligned}$$

$$= \frac{q}{r^2}\left(\hat{X} - \frac{A}{r}\right)\left(1 - \frac{3}{2}\left(\frac{|A|^2}{r^2} - 2\frac{A^\top \hat{X}}{r}\right)\right.$$
$$\left. + \frac{15}{8}\left(\frac{|A|^4}{r^4} - 4\frac{|A|^2}{r^3}A^\top \hat{X} + 4\frac{(A^\top \hat{X})^2}{r^2}\right) + \cdots\right)$$
$$= \frac{q}{r^2}\left(\hat{X} - \frac{A}{r}\right)\left(1 - \frac{3}{2}\frac{|A|^2}{r^2} + 3\frac{A^\top \hat{X}}{r}\right.$$
$$\left. + \frac{15}{8}\frac{|A|^4}{r^4} - \frac{15}{2}\frac{|A|^2 A^\top \hat{X}}{r^3} + \frac{15}{2}\frac{\hat{X}^\top A A^\top \hat{X}}{r^2} + \cdots\right) \quad (8)$$

Using the distributive law to multiply out the two parenthesized factors in equation (8), and grouping terms with the same negative power of r, we get

$$F(X) = \frac{q\hat{X}}{r^2} + \frac{q(3(A^\top \hat{X})\hat{X} - A)}{r^3}$$
$$+ \frac{q}{r^4}\left(-\frac{3}{2}|A|^2\hat{X} - 3AA^\top \hat{X} + \frac{15}{2}(\hat{X}^\top AA^\top \hat{X})\hat{X}\right) + \cdots . \quad (9)$$

In the case of several point charges, we sum over the point charges to get

$$F(X) = \frac{q_{total}\hat{X}}{r^2} + \frac{3(D \cdot \hat{X})\hat{X} - D}{r^3} + \frac{M\hat{X} + (\hat{X}^\top N\hat{X})\hat{X}}{r^4} + \cdots , \quad (10)$$

where the total charge $q_{total} = \sum_{i=1}^{n} q_i$, the dipole moment $D = \sum_{i=1}^{n} q_i A_i$, and the 3 by 3 matrices

$$\mathsf{M} = -3\sum_{i=1}^{n} q_i A_i A_i^\top - \left(\frac{3}{2}\sum_{i=1}^{n} q_i |A_i|^2\right) \mathsf{I}$$

and

$$\mathsf{N} = \frac{15}{2}\sum_{i=1}^{n} q_i A_i A_i^\top .$$

The first monopole term in equation (10) is linear in $\hat{X}$, the second dipole term is quadratic in $\hat{X}$, the third quadrupole term is cubic in $\hat{X}$, and so on. The monopole and dipole terms are as in equation 1.2.8 of [Sch94].

4.1 The monopole case

There are several cases, depending on which of these terms is the first non-zero one. Our goal is to find the radius r_m of a sphere containing all the critical points. If q_{total} is non-zero, we need to show that the magnitude $|q_{total}|/r^2$ of the monopole term is greater than the magnitude of the sum of all the rest of the terms, if $r > r_m$, for some appropriate r_m. If $q_{total} = 0$, we will instead use one of the higher order terms, as in sections 4.2 or 4.3.

Critical Points of the Electric Field from a Collection of Point Charges

We can bound the term in the line numbered (7) of the binomial series by

$$\left|\left(-\frac{3}{2}\right)\left(-\frac{5}{2}\right)\cdots\left(-\frac{2l+1}{2}\right)\frac{t^l}{l!}\right| = \frac{3}{2\times 1}\frac{5}{2\times 2}\cdots\frac{2l+1}{2\times l}|t|^l \leq \left|\frac{3}{2}t\right|^l.$$

Then this binomial series is dominated by a geometric series, and the sum of all its terms after the first is dominated by the sum of the geometric series:

$$\sum_{l=1}^{\infty}\left|\binom{-\frac{3}{2}}{l}t^l\right| \leq \sum_{l=1}^{\infty}\left|\frac{3}{2}t\right|^l = \frac{|\frac{3}{2}t|}{1-|\frac{3}{2}t|}.$$

For the ith point charge, let

$$t_i = \frac{|A_i|^2}{r^2} - 2\frac{A_i}{r}\cdot\hat{X}.$$

If we chose $r_m \geq 8k|A_i|$, for an as yet undetermined $k \geq 1$, and if $r > r_m$, then

$$|t_i| \leq \frac{|A_i|^2}{(8k|A_i|)^2} + 2\frac{|A_i|}{8k|A_i|}$$

$$\leq \frac{1}{64k^2} + \frac{2}{8k} \leq \frac{1}{64k} + \frac{2}{8k} = \frac{17}{64k}$$

so, using the fact that $k \geq 1$

$$\left|\sum_{l=1}^{\infty}\binom{-\frac{3}{2}}{l}t_i^l\right| \leq \frac{|\frac{3}{2}t_i|}{1-|\frac{3}{2}t_i|}$$

$$\leq \frac{\frac{51}{128k}}{1-\frac{51}{128k}} \leq \frac{\frac{51}{128k}}{1-\frac{51}{128}}$$

$$\leq \frac{\frac{51}{128k}}{\frac{77}{128}} = \frac{51}{77k}.$$

If we take $r_m = \max_i(8k|A_i|)$, for $k = \sum_{i=1}^n |q_i|/|\sum_{i=1}^n q_i| \geq 1$, sum over the contributions of all the the point charges, use the inequality $|B-C| \geq |B|-|C|$ for any two vectors B and C, and use the fact that $|\hat{X}| = 1$, we get, for $r > r_m$,

$$|F(X)| = \left|\sum_{i=1}^{n}\left(\frac{q_i}{r^2}\left(\hat{X}-\frac{A_i}{r}\right)(1+t_i)^{-3/2}\right)\right|$$

$$= \left|\sum_{i=1}^{n}\left(\frac{q_i}{r^2}\left(\hat{X}-\frac{A_i}{r}\right)\left(1+\sum_{l=1}^{\infty}\binom{-\frac{3}{2}}{l}t_i^l\right)\right)\right|$$

$$= \left|\sum_{i=1}^{n}\frac{q_i}{r^2}\hat{X} - \left(\sum_{i=1}^{n}\frac{q_i}{r^2}\frac{A_i}{r} - \sum_{i=1}^{n}\frac{q_i}{r^2}\left(\hat{X}-\frac{A_i}{r}\right)\sum_{l=1}^{\infty}\binom{-\frac{3}{2}}{l}t_i^l\right)\right|$$

$$\geq \left| \sum_{i=1}^{n} \frac{q_i}{r^2} \hat{X} \right| - \left| \sum_{i=1}^{n} \frac{q_i}{r^2} \frac{A_i}{r} - \sum_{i=1}^{n} \frac{q_i}{r^2} \left(\hat{X} - \frac{A_i}{r} \right) \sum_{l=1}^{\infty} \binom{-\frac{3}{2}}{l} t_i^l \right|$$

$$\geq \left| \sum_{i=1}^{n} \frac{q_i}{r^2} \right| - \left(\sum_{i=1}^{n} \frac{|q_i|}{r^2} \frac{|A_i|}{8k|A_i|} + \sum_{i=1}^{n} \frac{|q_i|}{r^2} \left(1 + \frac{|A_i|}{8k|A_i|} \right) \frac{51}{77k} \right)$$

$$= \frac{|\sum_{i=1}^{n} q_i|}{r^2} - \frac{\sum_{i=1}^{n} |q_i|}{r^2} \left(\frac{1}{8} + \frac{51}{77} \left(1 + \frac{1}{8k} \right) \right) \frac{|\sum_{i=1}^{n} q_i|}{\sum_{i=1}^{n} |q_i|}$$

$$= \frac{|\sum_{i=1}^{n} q_i|}{r^2} \left(1 - \left(\frac{1}{8} + \frac{51}{77} \left(1 + \frac{1}{8k} \right) \right) \right)$$

$$\geq \frac{|\sum_{i=1}^{n} q_i|}{r^2} \left(1 - \left(\frac{1}{8} + \frac{51}{77} \left(1 + \frac{1}{8} \right) \right) \right)$$

$$= \frac{|\sum_{i=1}^{n} q_i|}{r^2} \left(1 - \frac{536}{616} \right)$$

$$= \frac{|\sum_{i=1}^{n} q_i|}{r^2} \frac{80}{616}$$

$$> 0 ,$$

and thus there are no critical points outside the sphere of radius r_m.

4.2 The dipole case

If the total charge q_{total} is zero, as on a neutral molecule, but the dipole moment D is non-zero, we can proceed similarly, getting a lower bound on the dipole term, and showing that it dominates the remaining terms outside of a sufficiently large sphere.

To get the lower bound on the dipole term $(3(D \cdot \hat{X})\hat{X} - D)/r^3$, note that the points in the set $\{(D \cdot \hat{X})\hat{X}\}$ lie on a sphere of diameter $|D|$ centered at the point $\frac{1}{2}D$, since the point $(D \cdot \hat{X})\hat{X}$ is the foot of the perpendicular line from the point D to the line from the origin in the direction $\hat{X}$. (See figure 2 for a

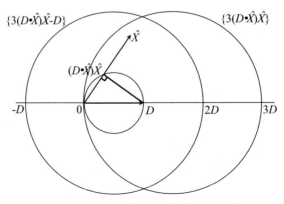

Fig. 2. Construction of the circle $\{3(D \cdot \hat{X})\hat{X} - D\}$ for all unit vectors $\hat{X}$

2D analog.) Thus the points of the form $3(D \cdot \hat{X})\hat{X}$ lie on a sphere of radius $\frac{3}{2}D$ centered at $\frac{3}{2}D$, and the translated points of the form $3(D \cdot \hat{X})\hat{X} - D$ lie on the translated sphere of radius $\frac{3}{2}D$ centered at $\frac{1}{2}D$. The closest point on this translated sphere to the origin is $-D$, and its distance to the origin is $|D|$. Thus a lower bound on the dipole term is $|D|/r^3$.

Since the dipole term arises from the terms $l = 0$ and $l = 1$ in the series (6) and (7), to show that the dipole dominates, we separate these terms from the remaining terms:

$$F(X) = \sum_{i=0}^{n} \left(\frac{q_i}{r^2} \left(\hat{X} - \frac{A_i}{r} \right) \left(1 - \frac{3}{2} \left(\frac{|A_i|^2}{r^2} - 2\frac{A_i^\top \hat{X}}{r} \right) + \sum_{l=2}^{\infty} \binom{-\frac{3}{2}}{l} t_i^l \right) \right)$$

$$= \frac{\sum_{i=0}^{n} q_i \hat{X}}{r^2} - \frac{\sum_{i=0}^{n} q_i A_i}{r^3} + \frac{3\left(\sum_{i=0}^{n} q_i A_i \cdot \hat{X}\right)\hat{X}}{r^3}$$

$$- \frac{\frac{3}{2}\sum_{i=0}^{n} q_i |A_i|^2 \hat{X}}{r^4} - \frac{3\sum_{i=0}^{n} q_i A_i A_i^\top \hat{X}}{r^4} + \frac{\frac{3}{2}\sum_{i=0}^{n} q_i |A_i|^2 A_i}{r^5}$$

$$+ \sum_{i=0}^{n} \frac{q_i}{r^2}\left(\hat{X} - \frac{A_i}{r} \right) \sum_{l=2}^{\infty} \binom{-\frac{3}{2}}{l} t_i^l . \qquad (11)$$

Of the seven terms in equation (11), the first is the monopole term, which we are assuming is zero, and the second and third are the dipole contribution, for which we just derived a lower bound. Similar to the monopole case, we can show that these dipole terms dominate the next three terms, because of their higher powers of r in the denominator, and also the last term, using the geometric series as above, if $r > r_m$, with

$$r_m = 36(\max_{i=1}^{n}|A_i|) \sum_{i=1}^{n} |q_i|/|D| . \qquad (12)$$

4.3 The quadrupole case

If both the total charge q_{total} and the dipole moment D are zero, and the quadrupole term is not zero, then the electric field decreases as $1/r^4$. Unlike the dipole case, we have not been able to derive a simple lower bound on the magnitude of the quadrupole term in equation (10). Instead, we apply the techniques in section 3, using the mean value theorem to get a lower bound.

We cover the unit sphere with a regular 2D latitude longitude (θ, ϕ) grid. Then, for each 2D grid cell, we find its 3D axis-aligned bounding box. For this box, we find bounds on the entries of the 3 by 3 matrix Q for the partial derivatives of the vector-valued function

$$G(\hat{X}) = \mathsf{M}\hat{X} + (\hat{X}^\top \mathsf{N}\hat{X})\hat{X} ,$$

with respect to the 3D Cartesian coordinates of $\hat{X}$. By the chain rule, the 3 by 2 partial derivative matrix P of $G(\hat{X}(\theta, \phi))$ with respect to θ and ϕ is the product of Q and the 3 by 2 Jacobian matrix J for the function $\hat{X}(\theta, \phi)$. We use the minimum and maximum on the 2D grid cell for the entries of J together with the lower and upper bounds on the 3D bounding box for the entries of Q, to find bounds on the entries of P. A 2D version of the method in section 3 can then give the bounds on the components of the vector $G(\hat{X})$, and thus a lower bound for $|G(\hat{X})|$ over the 2D cell. Let B be the minimum of this lower bound over all the 2D grid cells covering the sphere. (For the examples in this paper, we directly found $B > 0$ for the initial grid, so quadtree subdivision was not needed.) Then the norm of the quadrupole term is

$$\left| \frac{M\hat{X} + (\hat{X}^\mathsf{T} N \hat{X})\hat{X}}{r^4} \right| \geq \frac{B}{r^4} .$$

The vanishing monopole and dipole terms and the non-vanishing quadrupole term come from the $l = 0$, 1, and 2 terms of the series in (6) and (7), so, as for the dipole case, one can show that the quadrupole term dominates the sum of the six other terms from taking $l = 0$, 1, and 2 in equation (7), plus the sum

$$\sum_{i=1}^{n} \frac{q_i}{r^2} \left(\hat{X} - \frac{A_i}{r} \right) \sum_{l=3}^{\infty} \binom{-\frac{3}{2}}{l} t_i^l ,$$

when $r > r_m$, with

$$r_m = \max \left(6 \max_i |A_i|, \frac{61}{B} \sum_{i=1}^{n} |q_i| |A_i|^3 \right) . \tag{13}$$

We have not dealt with higher order cases, but they can be handled similarly, and are not normally needed.

5 Results

Figure 3 shows the saddle connectors between the critical points of an alanine residue isolated from a protein by cutting its peptide bonds. Standard partial charges making q_{total} zero were placed on the 10 atom centers, which are sources and sinks of the electric field. It took the 15 seconds on one processor of a dual 3.3 GHz Xeon workstation, to recursively search 457,328 octree cells to depth 44, and locate to 12 decimal place accuracy the 10 saddle points, using a sphere of size $r_m = 4$, known to contain all the critical points. However equation (12) gives $r_m = 238.4$, and then it took 14 hours to find the same critical points. Figure 4 shows the separation surface for this structure. These figures were drawn by a modification of the integration methods in [Th03] which uses the analytic gradient to integrate the separation surfaces and find the saddle connectors. Figure 5 shows the saddle connectors for a benzene molecule, with equal negative partial charges at each of its six carbon atoms,

Critical Points of the Electric Field from a Collection of Point Charges 111

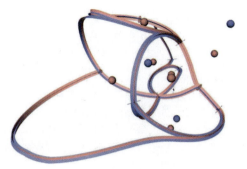

Fig. 3. The saddle connectors between the saddle points of an isolated alanine residue from a protein. The 10 atom centers are shown as pink spheres for sources (positive point charges), and blue spheres for sinks

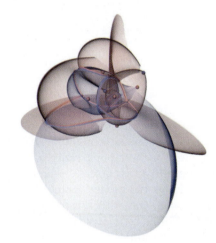

Fig. 4. The separation surfaces for the saddle points of an isolated alanine residue from a protein

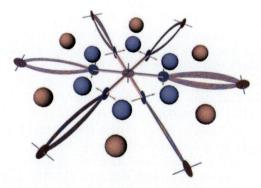

Fig. 5. The saddle connectors for a benzene molecule with 12 point charges

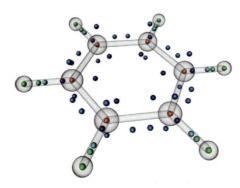

Fig. 6. A benzene molecule with 60 point charges

and positive charges at each of the six hydrogen atoms. Because of its symmetry, benzene has zero dipole moment as well as zero total charge, but it has a non-zero quadrupole moment.

The partial charge assignment at the atom positions gives only a rough approximation to the actual charge distribution generating the potential, and is certainly an oversimplification in the case of benzene, where the π orbitals from the carbon atoms lie above and below the plane of the 12 atom centers. We have obtained a more accurate quantum mechanical approximation of the electrostatic potential described in [St96], using 60 point charges. Figure 6 shows the arrangement of these point charges, which are not all in the same plane. It took 66 minutes to recursively search 23,290,456 octree cells to a depth of 44 and find the 193 critical points, using $r_m = 4$. However equation (13) gives $r_m = 9633$, and with this value, the algorithm ran for a week without terminating. So the estimates (12) and (13) for r_m are not very practical. Of the octree cells examined for $r_m = 4$, 14,833,200 were at level 8, while between 58,688 and 63,272 were at each of the levels 14 through 44. This shows that once the search locates the critical point neighborhood, the convergence is linear in the precision desired, and justifies our decision not to use a less robust method like Newton-Raphson iteration at the final stage. Figure 7 shows the saddle connectors for this collection of point charges, and figure 8 shows the separation surfaces. These figures are close to the corresponding figures in [Th03], which inspired the current work, but are different. In [Th03], the field was sampled on a 101^3 grid, and then trilinearly interpolated inside the grid cells. The critical points found for this interpolated field, using the methods of [Ma02], were 40 sinks, 12 sources, and 132 saddles. (Slightly different approximate numbers were reported in [Th03], but these are the counts gotten by rerunning that code on the same input.) We reran code in that paper on a 255^3 grid, and found 42 sinks, 12 sources, and 129 saddles, including 3 spurious saddles, one each in a face, an edge, and a vertex of the cube bounding the data volume. However, the analytic method described in the current paper found all 48 sinks and 12 sources at the input point charges,

Critical Points of the Electric Field from a Collection of Point Charges 113

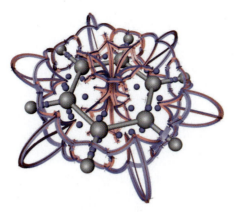

Fig. 7. The saddle connectors for a benzene molecule with 60 point charges

Fig. 8. The separation surfaces for a benzene molecule with 60 point charges

and 133 saddles. This includes 6 pairs of saddles and sinks, with the two so close together in each pair that they could not be resolved by the sampling method. Thus the analytic method is superior in this case.

Acknowledgments

This work was performed under the auspices of the U. S. Department of Energy by the University of California, Lawrence Livermore National Laboratory under contract number W-7405-ENG-48. We wish to thank Thomas Steinke for locating the data of the 60 point charges for benzene in an apparently lost old file, Daniel Laney, Hans-Christian Hege, and Frank Holzwarth for help with LaTeX, and Ajith Mascarenhas and Hans-Christian Hege for reading the manuscript and making suggestions. All visualizations in this paper have been created using AMIRA – a system for advanced visual data analysis [Am05] (see http://amira.zib.de/).

References

[deB00] de Berg, M., van Kreveld, M., Overmars, M., Schwarzkopf, O.: Computational Geometry: Algorithms and Applications, Second Edition, Springer, Berlin, (2000)

[Am05] Stalling, D., Westerhoff, M., Hege, H.-C.: Amira: A Highly Interactive System for Visual Data Analysis. In: Hansen, C., Johnson, C. (eds) The Visualization Handbook, Elsevier, 749–767 (2005)

[Ba90] Bader, R.F.W.: Atoms in Molecules: A Quantum Theory. Clarendon Press, Oxford (1990)

[Ja62] Jackson, J. D.: Classical Electrodynamics. Wiley, New York (1962)

[Ma02] Mann, S., Rockwood, A.: Computing Singularities of 3D Vector Fields with Geometric Algebra. In: Moorhead, R., Gross, M., Joy, K (eds) Proceedings of IEEE Visualization 2002, 283–289 (2002)

[Sch94] Scharf, G.: From Electrostatics to Optics: A Concise Electrodynamics Course. Springer-Verlag, Berlin (1994)

[St96] Stalling, D., Steinke, T.: Visualization of Vector Fields in Quantum Chemistry. ZIB Preprint SC-96-01. ftp://ftp.zib.de/pub/zib-publications/reports/SC-96-01.ps

[Th03] Theisel, H., Weinkauf, T., Hege, H.-C., Seidel, H.-P.: Saddle Connectors - An Approach to Visualizing the Topological Skeleton of Complex 3D Vector Fields. In: Turk, G, van Wijk, J.J., Moorhead, R. (eds) Proceedings of IEEE Visualization 2003, 225–232 (2003)

[Wiki] Wikipedia entry on spherical multipole moments. http://en.wikipedia.org/wiki/Spherical_multipole_moments

Visualizing global manifolds during the transition to chaos in the Lorenz system

Bernd Krauskopf[1], Hinke M Osinga[1], and Eusebius J Doedel[2]

[1] Bristol Centre for Applied Nonlinear Mathematics, Department of Engineering Mathematics, University of Bristol, Bristol BS8 1TR, United Kingdom
b.krauskopf@bristol.ac.uk, h.m.osinga@bristol.ac.uk
[2] Department of Computer Science, Concordia University, 1455 Boulevard de Maisonneuve O., Montréal, Québec H3G 1M8, Canada
doedel@encs.concordia.ca

Summary. If one wants to study the global dynamics of a given system, key components are the stable or unstable manifolds of invariant sets, such as equilibria and periodic orbits. Even in the simplest examples, these global manifolds must be approximated by means of numerical computations. We discuss an algorithm for computing global manifolds of vector fields that is decidedly geometric in nature. A two-dimensional manifold is built up as a collection of approximate geodesic level sets, i.e. topological smooth circles. Our method allows to visualize the resulting surface by making use of the geodesic parametrization.

As we show with the example of the Lorenz system, this is a big advantage when one wants to understand the geometry of complicated two-dimensional global manifolds. More precisely, for the standard system parameters, the origin of the Lorenz system has a two-dimensional stable manifold — called the Lorenz manifold — and the other two equilibria each have a two-dimensional unstable manifold. The intersections of these manifolds in the three-dimensional phase space form heteroclinic connections from the nontrivial equilibria to the origin. A parameter-dependent visualization of these manifolds clarifies the transition to chaos in the Lorenz system.

1 Introduction

We are concerned here with the problem of understanding the global behavior of a dynamical system that is defined by a set of ordinary differential equations. Written in the form of a vector field, the system takes the general form

$$\dot{\mathbf{x}} = \mathbf{f}(\mathbf{x}, \boldsymbol{\lambda}), \qquad (1)$$

where $\mathbf{x}$ is a point from an n-dimensional phase space $\mathbf{X}$, $\boldsymbol{\lambda}$ is a multidimensional parameter, and $\mathbf{f}$ is a sufficiently smooth (say, twice differentiable) vector-valued function. Indeed countless mathematical models arising in applications can be represented in this general framework; see, for example, [8, 21] as general entry points to the dynamical systems literature. As a specific example we consider below the well-known Lorenz system, which has the three-dimensional phase space $\mathbf{X} = \mathbb{R}^3$.

In order to understand the behavior of the system, one first considers the equilibria of (1), which are the points where $\mathbf{f}(\mathbf{x}, \boldsymbol{\lambda}) = \mathbf{0}$. An equilibrium is typically either an attractor, a repellor or a saddle point, depending on whether the eigenvalues of the linearization $\mathbf{Df}$ at the equilibrium have exclusively negative real parts, exclusively positive real parts, or are a mix of both, respectively. (Typical means here that there are no eigenvalues with zero real part.) The crucial role for organising the overall or global dynamics is played by the saddle points. Namely, a saddle point $\mathbf{x}_0$ comes with a stable manifold $W^s(\mathbf{x}_0)$ and an unstable manifold $W^u(\mathbf{x}_0)$, which are defined as the sets of all points in the phase space $\mathbf{X}$ that converge to $\mathbf{x}_0$ in forward and backward time, respectively. According to the Stable Manifold Theorem [17] these sets are actually smooth immersed manifolds that are tangent to (and of the same dimension as) the stable and unstable eigenspaces. The importance of these manifolds for the overall dynamics essentially lies in two facts. First of all, stable manifolds often act as boundaries of basins of attraction and, secondly, intersections of $W^s(\mathbf{x}_0)$ and $W^u(\mathbf{x}_0)$ are associated with chaotic dynamics.

In this paper we demonstrate the opportunities for the visualization of two-dimensional stable and unstable manifolds afforded by our approach to computing these global geometric objects. Our method, which generates a global manifold as a collection of geodesic level sets, is explained in Sect. 2. Throughout this paper we use the famous Lorenz system as an example. It is introduced in Sect. 3, where we also demonstrate how the knowledge of the geodesic level sets can be used to understand the geometry of the Lorenz manifold, that is, the stable manifold of the origin. We then take the visualization further when considering how the organisation of chaos varies with the Rayleigh number ϱ, which is a key parameter in the Lorenz system. Specifically, we show in Sect. 4 the transition of the Lorenz manifold through the first homoclinic bifurcation, which is also known as a homoclinic explosion point. We then show in Sect. 5 how the Lorenz manifold interacts with the unstable manifolds of saddle-periodic orbits and secondary equilibria of the Lorenz system when ϱ is increased up to the classic value of $\varrho = 28$. We summarize our findings in Sect. 6.

2 Global manifolds as a collection of geodesic level sets

The computation of global invariant manifolds in dynamical systems is an active field of research. The difficulty is that these objects are not given in the form of an implicit equation. Therefore, they need to be 'grown' by starting from local information, for example, near the saddle point. This is a nontrivial task already for manifolds of dimension two, which is the case considered here. Several methods are available today to compute global invariant manifolds (mostly of dimension two); see the recent survey [13]. Regardless of the choice of method, one is faced with the problem of visualizing the resulting manifolds in an efficient manner in order to extract the information on the global dynamics of the system.

Our method computes a global invariant manifold of, say, an equilibrium $\mathbf{x}_0$ in a very geometrical way, namely, by building it up step by step as a collection of geodesic level sets. Here, a geodesic level set is defined as the set of all points that lie at the same geodesic distance (arclength of the shortest connecting path on the manifold) from $\mathbf{x}_0$. Our method can be used, in principle, for computing manifolds of arbitrary dimension that are associated with arbitrary compact invariant objects, such as equilibria, periodic orbits, or higher-dimensional normally hyperbolic manifolds. It is presently implemented for two-dimensional manifolds in a phase space of any dimension.

We explain here how the method works for the important case of a two-dimensional stable manifold $W^s(\mathbf{x}_0)$ of a saddle point $\mathbf{x}_0$, where we follow [10, 5]. In this case the geodesic level sets that we seek are topological circles. The computation starts from a regular mesh M_0 on a small circle in the stable eigenspace of $\mathbf{x}_0$. The piecewise-linear curve C_0 through the mesh points in M_0 is the approximation of the first geodesic level set at some prescribed small distance from $\mathbf{x}_0$. Let us now assume that we already computed an approximation of $W^s(\mathbf{x}_0)$ up to the piecewise-linear curve C_i through mesh points M_i. The next step consists of finding a new approximate geodesic level set C_{i+1} at a prescribed distance Δ from C_i, which is computed pointwise. This step is illustrated in Fig. 1(a)–(c) for a given point $\mathbf{r} \in M_i$. We construct a plane $\mathcal{F}_\mathbf{r}$ through $\mathbf{r}$ (approximately) perpendicular to C_i. Then $W^s(\mathbf{x}_0) \cap \mathcal{F}_\mathbf{r}$ is locally a smooth one-dimensional curve through $\mathbf{r}$; see Fig. 1(a). By definition of $W^s(\mathbf{x}_0)$, points on this curve lie on orbits that converge to $\mathbf{x}_0$. Hence, to good approximation, these orbits pass through C_i; see Fig. 1(b). Therefore, we identify each point on $W^s(\mathbf{x}_0) \cap \mathcal{F}_\mathbf{r}$ locally near $\mathbf{r}$ as the start point of the orbit segment $\mathbf{u}(t)$, $0 \leq t \leq \tau$, that satisfies the boundary conditions

$$\mathbf{u}(0) = \mathbf{b}_\mathbf{r} \in \mathcal{F}_\mathbf{r},$$
$$\mathbf{u}(\tau) = \mathbf{q}_\mathbf{r} \in C_i,$$

where τ is the associated integration time. Starting from the trivial solution $\mathbf{u}(t) \equiv \mathbf{0}$, $0 \leq t \leq \tau$, with $\tau = 0$, the solution family of this two-point boundary value problem can be computed by continuation in the integration time τ. We stop the continuation as soon as $\mathbf{b}_\mathbf{r}$ lies at distance Δ from $\mathbf{r}$. If Δ is small enough, then such a point exists; see Fig. 1(c). In [5] this continuation procedure was implemented by using the pathfollowing and collocation routines of the package AUTO [2, 3].

When the mesh points $\mathbf{b}_\mathbf{r} \in M_{i+1}$ have been found for all $\mathbf{r} \in M_i$ it is checked whether neighbouring points are too close to or too far apart from each other. In the former case, a mesh point is deleted from M_{i+1}. In the latter case a new mesh point needs to be added. To ensure the overall accuracy of the

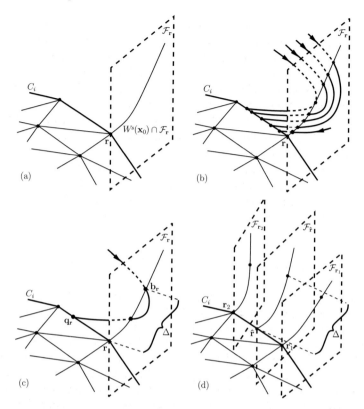

Fig. 1. Illustration of the main step of the algorithm. For each mesh point $\mathbf{r} \in C_i$ we consider the intersection of $W^s(\mathbf{x}_0)$ with a plane $\mathcal{F}_\mathbf{r}$ approximately perpendicular to C_i (a). To good approximation, points of $W^s(\mathbf{x}_0) \cap \mathcal{F}_\mathbf{r}$ lie on orbits through C_i (b). The new mesh point $\mathbf{b_r} \in \mathcal{F}_\mathbf{r}$ that we seek lies on such an orbit and exactly at distance Δ from $\mathbf{b_r}$ (c). Additional mesh points are added by using the same procedure for the point $\hat{\mathbf{r}} \in C_i$ that lies halfway in between two neighbouring mesh points $\mathbf{r}_1, \mathbf{r}_2 \in C_i$ (d)

mesh, we consider a new point $\hat{\mathbf{r}} \in C_i$ halfway in between the two underlying base points, say, $\mathbf{r}_1, \mathbf{r}_2 \in M_i$. We then apply the above continuation procedure to $\hat{\mathbf{r}}$ to find a new point in M_{i+1}; see Fig. 1(d).

Once a good representation of the next geodesic level set has been found in this way, we check whether Δ satisfies

$$\alpha_{\min} < \alpha_\mathbf{r} < \alpha_{\max},$$
$$(\Delta\alpha)_{\min} < \Delta \cdot \alpha_\mathbf{r} < (\Delta\alpha)_{\max},$$

for user-specified accuracy constants $\alpha_{\min}$, $\alpha_{\max}$, $(\Delta\alpha)_{\min}$ and $(\Delta\alpha)_{\max}$. Here, $\alpha_\mathbf{r}$ is the angle between the three corresponding mesh points in M_{i-1}, $\mathbf{r} \in M_i$ and $\mathbf{b_r} \in M_{i+1}$. These accuracy constraints are used to adapt Δ according to the local curvature of the manifold along geodesics [9, 10].

When the entire approximate geodesic level set C_{i+1} has been found, a triangulated band between C_i and C_{i+1} is added to the mesh representation of the surface. The computation stops when a pre-specified geodesic distance has been reached. A particular feature of our method is that the control over the mesh quality guarantees correctness: any mesh that is computed up to a prescribed geodesic distance with sufficient accuracy (as determined by the user) lies within an ε-neighbourhood of the global manifold in the Hausdorff metric; see [10] for details and the proof. What is more, the result of a computation is a natural and geometric representation of the manifold in terms of a *geodesic mesh*, which consists locally near each mesh point of a near-perpendicular intersection of approximate geodesic level sets and approximate geodesics. It is this property that we exploit in the visualizations presented here. We used Geomview [19] for the rendering of the manifolds.

3 Visualizing the Lorenz manifold

The Lorenz system still fascinates many people because of the simplicity of the equations that generate such complicated dynamics on the famous butterfly attractor. It is arguably the most famous dynamical system exhibiting chaotic dynamics. When Lorenz examined the behaviour of Rayleigh-Bénard convection in 1963, he found that for Rayleigh numbers well after the critical value of onset of convection the behaviour of the fluid is aperiodic. Since the motion is nevertheless bounded, he could prove that an attractor must exist along with infinitely many unstable periodic orbits; see [14]. In his paper Lorenz uses extremely simplified equations by considering only one mode for the velocity and two modes for the temperature of the fluid. This results in the system that is now know as the Lorenz system, which can be written in the form (1) as the vector field

$$\begin{cases} \dot{x} = \sigma(y-x), \\ \dot{y} = \varrho x - y - xz, \\ \dot{z} = xy - \beta z. \end{cases} \quad (2)$$

The Lorenz system has the three-dimensional phase space $\mathbb{R}^3$, and its main parameter is the Rayleigh number ϱ; note that system (2) is rescaled such that the onset of convection occurs at $\varrho = 1$. Lorenz provides physically relevant values for the parameters in [14], namely, the now classic choice $\varrho = 28$, $\sigma = 10$ and $\beta = \frac{8}{3}$, for which Lorenz found sensitive dependence on the initial condition and the well-known butterfly or Lorenz attractor [14].

We are interested here in the stable manifold $W^s(\mathbf{0})$ of the origin, which is a saddle point with one unstable and two stable eigenvalues. Hence, $W^s(\mathbf{0})$ is a two-dimensional smooth surface. Geometrically, it divides points that initially go to one or the other 'wing' of the Lorenz attractor. More generally, $W^s(\mathbf{0})$ organises the chaotic dynamics of the Lorenz system in a global way, which is why we refer to it as the *Lorenz manifold*; see also [15, 11]. The visualization of the Lorenz manifold is quite a challenge. Topologically, any finite piece of $W^s(\mathbf{0})$ is a disk, but one whose embedding in $\mathbb{R}^3$ becomes increasingly complicated geometrically. The first images of $W^s(\mathbf{0})$ are the hand-drawn sketches by Perelló from 1979 in [18]. His work formed the basis for the drawings of Shaw in [1].

Figure 2 shows the Lorenz manifold $W^s(\mathbf{0})$ computed up to geodesic distance 151.75; the origin $\mathbf{0}$ is in the middle of the images and the vertical axis is the z-axis, which is invariant under (2). Note further that, due to the symmetry $(x, y) \mapsto (-x, -y)$ of the Lorenz system, $W^s(\mathbf{0})$ is symmetric under a rotation by π about the z-axis. The Lorenz manifold is rendered transparent in Fig. 2 so that its 'internal' structure can be seen. The first feature one notices is the main helix of $W^s(\mathbf{0})$ around the positive z-axis. Notice also the two secondary (and symmetrically related) helices near the main helix. These helices arise because $W^s(\mathbf{0})$ spirals around two smooth symmetrical curves (the

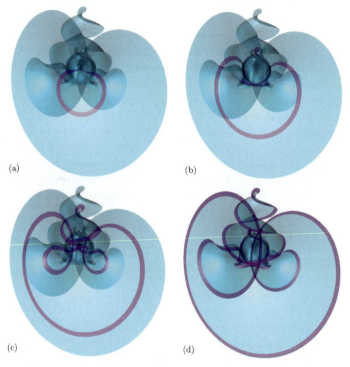

Fig. 2. The Lorenz manifold $W^s(\mathbf{0})$ of system (2) for the classic value of $\varrho = 28$ computed up to geodesic distance 151.75 and rendered transparent. To bring out its complicated geometry, a contrastingly coloured transparent geodesic band is moved over the surface; shown is the band covering geodesic distances 38.75–46.75 (a), 74.75–82.75 (b), 110.75–118.75 (c), and 144.75–151.75 (d)

one-dimensional stable manifolds of a symmetric pair of nontrivial saddle points; not shown in Fig. 2) in opposite directions.

Even when it is rendered transparent it is not easy to understand the intricate geometry of $W^s(\mathbf{0})$. To help with the visualization we move a geodesic band of a contrasting colour over the surface; it is shown in four different positions in Fig. 2. A band near the origin is small and almost perfectly round. For increasing geodesic distance from the origin the band starts to pick up the spiralling along the main helix; see Fig. 2(a). It then spirals more near the origin and simultaneously moves up the main helix; see Fig. 2(b) and (c). For even larger geodesic distances this results in the creation of the two secondary helices on the band, as is illustrated in Fig. 2(d) with the outer most band of the computed surface.

It is important to realise that each geodesic band is unknotted. This corresponds to the fact that the Lorenz manifold itself is topologically simply a disk. The geodesic mesh representation of $W^s(\mathbf{0})$ that we compute can be interpreted as an illustration of an (unknown) smooth map that embeds the standard disk into $\mathbb{R}^3$. As is clear from Fig. 2, such an embedding can be very complicated. We finally remark that the geodesic mesh computed by our method translates naturally into crochet instruction, which allowed us to make a real model of the Lorenz manifold; see [16] for details and images of the crocheted Lorenz manifold.

4 Transition through the homoclinic explosion

At the homoclinic explosion at $\varrho_{\text{hom}} \approx 13.9162$ each branch of the one-dimensional unstable manifold of the origin forms a homoclinic loop after a single rotation around one of the two secondary equilibria

$$p^\pm = (\pm\sqrt{\beta(\varrho-1)}, \pm\sqrt{\beta(\varrho-1)}, \varrho-1).$$

As is well known, infinitely many periodic orbits of the Lorenz system are created in this bifurcation, including a pair of saddle periodic orbits $\Gamma^\pm$; see [20, 4]. Due to the symmetry of rotation by π around the z-axis, p^+ and p^-, and Γ^+ and Γ^-, are each other's symmetric counterparts.

We are interested here in visualizing how this homoclinic bifurcation influences the geometry of $W^s(\mathbf{0})$. Figure 3 shows $W^s(\mathbf{0})$ just before, and just after the homoclinic explosion, namely at $\varrho = 13$ and at $\varrho = 15$, respectively. The manifold $W^s(\mathbf{0})$ is now shown up to geodesic distance 100 and again transparent. To highlight its exact geometric structure the outermost geodesic band is shown in a contrasting colour.

Figure 3(a) illustrates the situation for $\varrho < \varrho_{\text{hom}}$, namely for $\varrho = 13$. In this case the Lorenz manifold wraps around $p^\pm$ once, returning back near $\mathbf{0}$ at a negative z-value, after which it folds down (towards negative z) such that it lies practically flat against the lower half-disk of $W^s(\mathbf{0})$. At the same

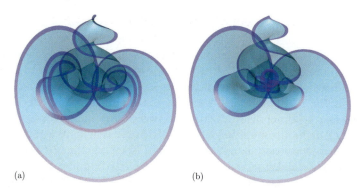

(a) (b)

Fig. 3. The Lorenz manifold $W^s(\mathbf{0})$ computed up to geodesic distance 100 for $\varrho = 13$ (a) and for $\varrho = 15$ (b). To visualize how $W^s(\mathbf{0})$ changes in the transition through the homoclinic explosion at $\varrho_{\text{hom}} \approx 13.9162$, we show a contrastingly coloured outer band (geodesic level sets at distances 97–100) and render the surfaces transparent

time, there is the helix along the positive z-axis. The visualization in Fig. 3 allows one to follow the coloured outer band of $W^s(\mathbf{0})$ to obtain an idea of the overall geometry of the surface. To this end, consider starting at the lowest point $(0, 0, -100)$ and moving to the left. The outer band turns up and around before making a relatively sharp turn slightly up and back down towards the negative z-axis. With some practice, one can observe that the band continues further to the right and starts winding its way up the helix until it reaches $(0, 0, 100)$. When moving right from $(0, 0, -100)$ the symmetrical behaviour can be observed.

The situation for $\varrho > \varrho_{\text{hom}}$ is visualized in Fig. 3(b) for $\varrho = 15$. As one follows the outer band to the left from the lowest point $(0, 0, -100)$ the initial behavior is as before; indeed this first part of the outer band is virtually identical to that shown in Fig. 3(a). However, where the band in Fig. 3(a) suddenly turns slightly up and back down towards the negative z-axis, in Fig. 3(b) the relatively sharp turn is in the exact opposite direction and the band never passes near the negative z-axis before reaching $(0, 0, 100)$.

It is important to realize that this dramatic change in the geometry of the geodesic level sets is not due to the relatively large gap between the two ϱ-values, 13 and 15, before and after the homoclinic bifurcation. The switch between passing by the negative z-axis again or not is sudden and immediate and a result of the existence of the homoclinic loop at $\varrho_{\text{hom}} \approx 13.9162$. In fact, exactly at the homoclinic explosion, the Lorenz manifold returns to and 'closes up' in a non-smooth way along a special curve known as the strong stable manifold of the origin; see [4] for details.

5 Intersections of two-dimensional manifolds

In the homoclinic explosion two periodic orbits $\Gamma^{\pm}$ are created, which are of saddle type. Therefore, they come with two-dimensional unstable manifolds $W^u(\Gamma^{\pm})$. Due to the symmetry of the Lorenz system it suffices to compute only, say, $W^u(\Gamma^+)$. As ϱ is increased the periodic orbits $\Gamma^{\pm}$ shrink down in size. The Lorenz attractor is created in a heteroclinic bifurcation between the origin and $\Gamma^{\pm}$, which takes place at $\varrho_{\text{het}} \approx 24.0579$. Finally, the periodic orbits disappear in a subcritical Hopf bifurcation of the secondary equilibria $p^{\pm}$ at $\varrho_{\text{H}} \approx 24.7368$. As a consequence, $p^{\pm}$ lose their stability and become saddles with two-dimensional unstable manifolds $W^u(p^{\pm})$. The manifolds $W^u(p^+)$ and $W^u(\Gamma^+)$ consist of infinitely many layers that are extremely close together. By identifying these layers one obtains a branched surface known as the template of the Lorenz attractor, which can be used to describe the chaotic dynamics in the Lorenz system [6, 7, 24].

The key in the transition from the homoclinic explosion at ϱ_{hom} to the classic situation for $\varrho = 28$ is to understand how the Lorenz manifold intersects $W^u(p^{\pm})$ and $W^u(\Gamma^{\pm})$, respectively. To compute a suitable first part of $W^u(p^+)$ or $W^u(\Gamma^+)$, which effectively represents the template, we continue an orbit segment that starts on a fixed vector in the unstable eigenspace of p^+ and ends in the section $\{z = \varrho - 1\}$ either near p^+ or near p^-. This continuation is done with the package AUTO [2, 3]; see [4, 12] for a more detailed description of this method.

Figure 4 shows the Lorenz manifold $W^s(\mathbf{0})$ for four values of ϱ as it intersects the unstable manifolds $W^u(\Gamma^{\pm})$ (panels (a)–(c)) and $W^u(p^{\pm})$ (panel (d)). In these images $W^s(\mathbf{0})$ is rendered transparent, while $W^u(\Gamma^{\pm})$ and $W^u(p^{\pm})$ are shown as solid surfaces. In this way, one gets an impression of the intersection of the two surfaces. The viewpoint in Fig. 4 is fixed and chosen exactly as for Fig. 3; compare Fig. 4(a) with Fig. 3(b). In particular, the vertical axis is again the z-axis. Figure 4(a)–(c) shows $W^u(\Gamma^{\pm})$ for $\varrho = 15$, $\varrho = 19$ and $\varrho = 23$, respectively. Notice how $W^u(\Gamma^{\pm})$ grows while the periodic orbits $\Gamma^{\pm}$ actually shrink. They surround the secondary equilibria $p^{\pm}$, which are attractors. Finally, in Fig. 4(d) the periodic orbits $\Gamma^{\pm}$ are gone and the image shows the manifolds $W^u(p^{\pm})$ of the equilibria $p^{\pm}$, which are now saddle points. We remark that this transition from $W^u(\Gamma^{\pm})$ to $W^u(p^{\pm})$ is smooth (i.e., is not noticeable on the level of these two-dimensional surfaces). Notice further that, as ϱ increases, the amount of spiralling of the helix around the positive z-axis decreases.

Keeping in mind that the red unstable manifolds $W^u(\Gamma^{\pm})$ and $W^u(p^{\pm})$ in Fig. 4 actually consist of infinitely many sheets, we conclude that they have infinitely many intersection curves with the Lorenz manifold $W^s(\mathbf{0})$. These intersection curves are structurally stable heteroclinic orbits that connect the origin with $\Gamma^{\pm}$ and $p^{\pm}$, respectively. In fact, as is shown in [4], there is a ϱ-dependent hierarchy of heteroclinic orbits that can be described in terms of symbolic dynamics.

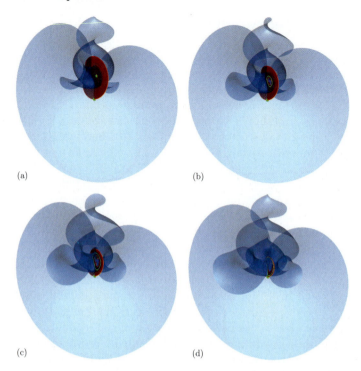

Fig. 4. Transparent renderings of the Lorenz manifold $W^s(\mathbf{0})$ computed up to geodesic distance 100 as it intersects the unstable manifolds $W^u(\Gamma^\pm)$ and $W^u(p^\pm)$, respectively, for $\varrho = 15$ (a), for $\varrho = 19$ (b), for $\varrho = 23$ (c), and for $\varrho = 28$ (d).

6 Conclusions

The computation of global stable and unstable manifolds is an important tool when one wants to obtain an understanding of how the dynamics of a vector field is organized. Particularly in the presence of chaotic dynamics, the geometry of global invariant manifolds can be very complicated, so that an appropriate visualization is a necessity.

We used the example of the Lorenz system to demonstrate how the geodesic mesh representation that is the result of our method in [5, 10] can be used to visualize two-dimensional global manifolds. First of all, we demonstrated that moving a contrastingly coloured geodesic band over the Lorenz manifold (the stable manifold of the origin) adds an new dimension to the visualization of its intricate geometry. We then showed how the Lorenz manifold changes during the homoclinic bifurcation that gives rise to the complicated dynamics of the Lorenz system and how it intersects the unstable manifolds of the bifurcating periodic orbits and secondary equilibria.

References

1. Abraham, R.H., Shaw, C.D.: Dynamics — The geometry of behavior, Part three: global behavior. Aerial Press, Santa Cruz (1985)
2. Doedel, E.J.: AUTO, a program for the automatic bifurcation analysis of autonomous systems. Congr. Numer., 30, 265–384 (1981)
3. Doedel, E.J., Champneys, A.R., Fairgrieve, T.F., Kuznetsov, Yu.A., Sandstede, B., Wang, X.J.: AUTO97: Continuation and bifurcation software for ordinary differential equations. available via http://cmvl.cs.concordia.ca/auto/ (1997) (accessed November 2006)
4. Doedel, E.J., Krauskopf, B., Osinga, H.M.: Global bifurcations of the Lorenz manifold. Nonlinearity, 19(12), 2947–2972 (2006)
5. England, J.P., Krauskopf, B., Osinga, H.M.: Computing two-dimensional global invariant manifolds in slow-fast systems. Int. J. Bifurcation and Chaos, 17(3), 805–822 (2007)
6. Ghrist, R., Holmes, P.J., Sullivan, M.C.: Knots and links in three-dimensional flows. Lecture Notes in Mathematics 1654, Springer, Berlin (1997)
7. Gilmore, R., Lefranc, M.: The topology of chaos: Alice in stretch and squeezeland. Wiley-Interscience, New York (2004)
8. Guckenheimer, J., Holmes, P.: Nonlinear Oscillations, Dynamical Systems and Bifurcations of Vector Fields. 2nd Printing, Springer-Verlag, New York (1986)
9. Hobson, D.: An efficient method for computing invariant manifolds of planar maps. J. Comput. Phys. 104(1), 14–22 (1993)
10. Krauskopf, B., Osinga, H.M.: Computing geodesic level sets on global (un)stable manifolds of vector fields. SIAM J. Appl. Dyn. Sys., 4(2), 546–569 (2003)
11. Krauskopf, B., Osinga, H.M.: The Lorenz manifold as a collection of geodesic level sets. Nonlinearity 17(1), C1–C6 (2004)
12. Krauskopf, B., Osinga, H.M.: Computing invariant manifolds via the continuation of orbit segments. In: Krauskopf, B., Osinga, H.M., Galán-Vioque, J. (Eds.): Numerical continuation methods for dynamical systems. Springer Complexity: Understanding Complex Systems, Springer, Berlin, pp. 117–154 (2007)
13. Krauskopf, B., Osinga, H.M., Doedel, E.J., Henderson, M.E., Guckenheimer, J., Vladimirsky, A., Dellnitz, M., Junge, O.: A survey of methods for computing (un)stable manifolds of vector fields. Int. J. Bifurcation and Chaos 15(3), 763–791 (2005)
14. Lorenz, E.N.: Deterministic nonperiodic flows. J. Atmospheric Sci., 20, 130–141 (1963)
15. Osinga, H.M., Krauskopf, B.: Visualizing the structure of chaos in the Lorenz system. Computers and Graphics, 26(5), 815–823 (2002)
16. Osinga, H.M., Krauskopf, B.: Crocheting the Lorenz manifold. Math. Intelligencer, 26(4), 25–37 (2004)
17. Palis, J., de Melo, W.: Geometric Theory of Dynamical Systems. Springer, New York (1982)
18. Perelló, C.: Intertwining invariant manifolds and Lorenz attractor. In: Global theory of dynamical systems (Proc. Internat. Conf., Northwestern Univ., Evanston, Ill., 1979). Lecture Notes in Math. 819, Springer, Berlin, pp. 375–378 (1979)
19. Phillips, M., Levy, S., Munzner, T.: Geomview: An Interactive Geometry Viewer. Notices of the American Mathematical Society, 40: 985–988 (1993); This software and the accompanying manual are available at http://www.geom.uiuc.edu/ (accessed November 2006)

20. Sparrow, C.: The Lorenz Equations: Bifurcations, Chaos and Strange Attractors. Appl. Math. Sci. No. 41, Springer, New York (1982)
21. Strogatz, S.: Nonlinear Dynamics and Chaos. Addison Wesley, Boston, MA (1994)
22. Tucker, W.: The Lorenz attractor exists. C. R. Acad. Sci. Paris Sér. I Math., 328(12), 1197–1202 (1999)
23. Viana, M.: What's new on Lorenz strange attractors? Math. Intelligencer, 22(3), 6–19 (2000)
24. Williams, R.F.: The universal templates of Ghrist. Bull. AMS, 35(2), 145–156 (1998)

Streamline and Vortex Line Analysis of the Vortex Breakdown in a Confined Cylinder Flow

Markus Rütten[1] and Gert Böhme[2]

[1] German Aerospace Center (DLR), Institute of Aerodynamics and Flow Technology markus.ruetten@dlr.de
[2] Institute of Mechanics, Helmut-Schmidt-University, gert.boehme@hsu-hh.de

Summary. The vortex breakdown phenomenon occurring in a rotating flow within a closed cylinder is still a challenging research field. In particular the goal to describe all significant order parameters of vortex breakdown is not reached. For further insight the viscous and laminar Newtonian flow inside a cylinder with a rotating lid has been calculated by solving the full Navier Stokes equations. During the simulation the rotational speed of the lid has been increased, which causes a gradual transition of the internal flow field topology. Starting from a flow field without any reversed flow at the vortex axis the vortex breakdown phenomenon develops indicated by one or more vortex breakdown bubbles. A phenomenological description of the vortex breakdown process is given by applying a topological analysis to the flow field, which illustrates the main flow structures, their behaviour and changes. By visualization of critical points, at which the velocity magnitude vanishes, the topological flow structure change of the velocity field becomes obvious. Additionally their associated separatrices are integrated into the field, which allows to illustrate the shape of the vortex breakdown bubbles. In particular the spherical shape of the first appearing breakdown bubble leads to the idea to introduce a streamfunction, which describes the spherical breakdown bubble approximately. Applying a Taylor expansion of the velocity field leads to an analytical description of the local streamline topology nearby one critical point of a breakdown bubble. The interpretation of the appendant differential equations allows a deeper insight into the dynamical behaviour of the breakdown phenomenon and its main enforcing parameters. The paper presents the results of a local streamline and vortex line topology analysis, especially the dynamical relation between the velocity and vorticity field in regard to the topological structure of the vortex breakdown phenomenon in the lid driven cylinder.

1 Introduction

This paper deals with the flow phenomenon called vortex breakdown. The term vortex breakdown describes the abrupt change of a nearly columnar vortex into a much larger flow structure: The former small streamsurface closely surrounding the vortex axis increases suddenly, then three types of new vortical flow structures can occur. For relatively low Reynolds numbers huge bubble like flow structures, the vortex breakdown bubbles, can be observed. Those bubbles consist of a recirculation zone with backward directed flow. In the case of a medium Reynolds number the columnar vortex can decompose into one or more meandering and spiraling sub vortices, also this vortex breakdown type shows a reversed flow. A third type of vortex breakdown is characterized by the fact that the former columnar vortex is spread into a highly turbulent conical flow region, which almost consists of small scale turbulent flow structures. The last breakdown type can only occur in turbulent flows at a high Reynolds number. The multifaceted appearance of the phenomenon vortex breakdown has often been described since it occurs either in nature, i.e. in tornadoes, or in many technical applications like delta wing flows, the flow in burning chambers or in mixers. Beside many specific publications related to certain aspects of the vortex breakdown the reviews of Leibovich [20], Escudier [13], Delery [11] and Sarpkaya [25] provide a good overview of the difficult issue. Although the topic is a long time research object, until now, there is no complete theoretical description of all aspects of the flow phenomenon and its development. Often only certain aspects of the vortex breakdown are regarded in relation to a special class of technical applications.

The common overall goal is to control this flow phenomenon in the sense of either preventing a vortex breakdown for instance in regard to aircraft applications and hydrodynamic turbines or enforcing it to aim a better mixing of fluids, as in burning chambers or in pharmaceutical production processes. Bearing this in mind there is the interest to find the significant physical parameters of the vortex breakdown phenomenon. Unfortunately in almost all applications the manifold of the related geometrical and physical parameters prevents an analytical description or formulation of an exact theory for the vortex breakdown. From the scientific point of view it is not convenient to consider the flow of complex configurations keeping the goal in mind to catch all order parameters without understanding simpler configuration. As a consequence strong simplifications of the geometrical complexity or idealisations are necessary. Hence, we concentrate on the Newtonian vortical flow and its flow structure change inside a closed cylinder with an instationary rotating lid. This type of cylinder flow depends on two parameters only: the aspect ratio of the cylinder and the Reynolds number. By choosing a certain cylinder height-to-diameter ratio the problem reduces to a one parameter case. The remaining parameter Reynolds number Re is made up of the radius R of the cylinder which is, at the same time, the radius of the rotating lid, the angular velocity Ω and the kinematic viscosity ν of the liquid:

$$Re = \frac{R^2 \Omega}{\nu}. \tag{1}$$

For this cylinder configuration two or three vortex breakdown bubbles can occur.

In the view of numerical parameter studies the torsionally driven cylindrical flow has further advantages: Geometrically caused there is one defined vortex axis for both cases, the flow with and without breakdown bubbles. This is the symmetry axis of the cylinder. Therefore, it is well known, where the topological flow structure change in form of the vortex breakdown process will begin. Hence it is possible to resolve the interesting region accurately regarding the needs of CFD. Furthermore, the underlying symmetry can be used in analysing the flow field allowing specific assumptions and simplifications.

Bearing in mind theses advantages Hussain et al. [18], Goldshik et al. [15], Spohn et al. [28] and Lugt [22] have done a lot of experimental work to describe the fundamental flow structure and its developing inside such a configuration. By injecting dye into the cylinder they could observe the evolving of one or more local stationary vortex breakdown bubbles with included backward directed flow. Escudier [12] systematically investigated the flow behaviour in such cylinders considering different aspect ratios and Reynolds numbers and gave a detailed overview about the bubble like vortex breakdown structures and their appearance.

One important feature of those breakdown bubbles is the formation of two stagnation respectively critical points. Following the definition of Chong [7] these are points in the free flow, at which the velocity vector vanishes. Separatrices, starting from critical points, are limiting the bubble type vortex breakdown and are describing the shape of the breakdown bubble. In the case of a steady flow with a geometrically stable breakdown bubble the separatrices are connecting both critical points, in the instationary lid driven flow the bubbles are increasing or decreasing, then the separatrices, starting from one critical point, are passing the other critical point nearby. Inside the bubble a roughly toroidal flow establishes with a backwards directed velocity component in the center of rotation. In this paper we will concentrate on Newtonian flows only. Böhme et al. [4] could show that the vortex breakdown phenomenon can also appear in shear-thinning non-Newtonian flows. Their numerical calculations could match the experiments pretty well. However both, numerical simulations and experiments reveal major differences to Newtonian flows. For instance, at $H/R = 2$ no further breakdown bubble was found with a shear-thinning non-Newtonian liquid although the observed breakdown bubble had nearly the same shape as the observed primary bubble in the Newtonian flow case.

We have organized the material presented here as follows: First the numerical experiment with its parameters is described. Then an impression of the main flow behaviour is given. The topological change of the principle flow structure, the occurrence and decay of the vortex breakdown bubbles is visualised by streamlines and isosurfaces. Therewith, a phenomenological description of the vortex breakdown follows. The next part describes the topological analysis which starts with detection and visualisation of critical points and calculating their separatrices. This approach allows to reduce the flow field information strongly and to concentrate on the main flow pattern in the visualization. In particular the shape of the separatrices are interesting for the further analysis.

Thereafter, to get a further link to a physical explanation of vortex breakdown the vortex lines nearby the vortex axis are calculated. Their behaviour will be compared to that of the streamlines. For the present paper the streamline and vortex line topologies are the focal point. In contrast to Brons et al. [5], who are applying dynamical system approaches for analysing the bifurcation behaviour of degenerated critical points and their separatrices, differential equations are formulated for the velocity field nearby non-degenerated critical points. By considering symmetry conditions of the global flow field and concentrating on the shape of one breakdown bubble we are able to simplify the set of equations and derive the streamfunction for the vortex breakdown bubble. This enables us to analyse the dynamics of the related terms.

2 Numerical Experiment of the Lid Driven Cylinder Flow

For the purpose of numerical flow simulation a closed cylinder with a height-to-diameter ratio of one ($H/R = 2$) was considered. This geometry has the advantage that two vortex breakdown bubbles can be expected, and the cylinder is relatively compact. This could be used to enlarge the spatial resolution of the grid without exceeding the number of points. Symmetrical considerations would allow to reduce the grid to a pie slice and to apply cylindrical symmetric boundary conditions, decreasing the amount of grid points furthermore. However, we decided to calculate the whole cylinder geometry. The disadvantage to spend more points than necessary has been compensated by both the advantage of an easier post-processing, in particular for the streamline integration, and avoiding the implementation of periodic boundary conditions. For grid generation we choose a cylindrical coordinate system with origin at the bottom. The grid has 80 points in radial direction, and 120 points in circumferential direction and in axial direction. At the walls of the cylinder the non-slip boundary condition was assumed. Thus, the velocity vanishes at the fixed walls, whereas at the rotating lid on top a solid body rotation was implemented.

To solve the Navier-Stokes equations and to carry out the laminar flow simulation the incompressible DLR code THETA was used. This second order finite volume code works on unstructured and hybrid grids, supporting tetrahedral, prismatic, pyramidal and hexahedral elements. Due to a dual grid technique and an edge-based data structure, the flow solver is completely independent of the cell types of the primary hybrid grid. The flow variables are stored at the centers of the dual grid at the vertices of the primary grid. A variant of the projection method of Chorin which enables a sequential calculation of velocity and pressure at each time step was used for time dependent calculations. In order to solve the momentum equations a quadratic upwind differential scheme (QUDS) was applied, the pressure part was solved by a generalized minimal residual method (GMRES). Additionally an agglomeration multi-grid method with integrated Jacobi smoothing steps was utilised.

Unsteady simulations were carried out, starting from a converged steady solution. Initial Reynolds number was 1200. The kinematic viscosity was set to a value of $1.4 \cdot 10^{-4} m^2 s^{-1}$, which represents a thin mineral oil. The starting angular velocity of the lid was $0.672 s^{-1}$ and as reference length a diameter of $1m$ was selected. With these parameters the flow was calculated until the residual was decreased to a value of 10^{-6}. The initial solution for the unsteady calculation represents a stable vortical flow structure. Unsteady flow simulations were performed with an angular acceleration of the lid of $2.24 \cdot 10^{-4} s^{-2}$ and a physical time step of $1s$. Setting these relatively small simulation parameters provided sufficient relaxation time for a very gentle and smooth flow structure change, so all significant flow phenomena could be resolved. With gradually increasing of the angular velocity of the lid first the domain of one breakdown bubble was reached at a Reynolds number of 1450. Later at a Reynolds number of 1800 the domain with two bubbles was reached. In the following at a Reynolds number value of 2100 the second bubble vanishes, and finally at a Reynolds number of 2950 the breakdown phenomenon disappeared completely.

3 Phenomenological Description Of Vortex Breakdown

In the beginning of the simulation no vortex breakdown bubble is present. Then the angular speed of the lid has been linearly increased. After exceeding the Reynolds number of 1450 the first bubble occurs.

In figure 1 the increased revolution of the lid has enforced the first vortex breakdown bubble. In 1(a) the vortex breakdown structure is indicated by streamlines, which are seeded at a cylinder height of $0.1m$ above the bottom. They are arranged on a small cycle with a diameter of $0.01m$. As upper integration limit a height of $0.9m$ was chosen to concentrate on the vortex core by avoiding multiple line integration in the same region. This arrangement of streamlines forms a streamtube that surrounds the vortex axis. The shape change of the small streamtube can indicate the vortex breakdown bubble which is additionally illustrated by an isosurface of the negative axial velocity component. When observing the isosurfaces we have to remind that the outer rolling in part encloses the back flow region, so the vortex structures are somewhat larger and rounder than visualised by the axial velocity component. As complement in figure 1(b) the meridional flow field is visualized by a line integration convolution (LIC) texture [6] of the meridional velocity field. Here, the breakdown bubble is indicated by the toroidal texture structure.

Further increasing of the angular velocity of the lid causes a spatial enlargement of the primary breakdown bubble until the second bubble arises. In figure 2(a) the streamlines show a relatively big cylindrical structure above

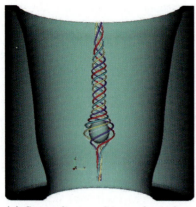 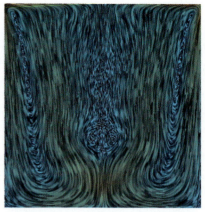

(a) Streamlines and isosurface of negative axial velocity (-1e-4 m/s), Reynolds number 1600

(b) LIC texture of the meridional velocity field, Reynolds number 1600

Fig. 1. Flow structure visualization by streamlines and LIC Textures

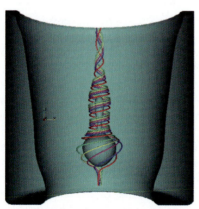 

(a) Streamlines and isosurface of negative axial velocity (-1e-4 m/s), Reynolds number 2000

(b) LIC texture of the meridional velocity field, Reynolds number 2000

Fig. 2. Visualization of the fused breakdown bubbles

the nearly spherical region of the primary bubble. Inside this upper structure the secondary bubble develops, which subsequently fuses with the primary bubble. Inside the fused bubbles the former toroidal structure of the primary breakdown bubble has begun to deform to a goblet like form, which reveals the texture in figure 2(b).

After the fusion of both bubbles the decaying process of the vortex breakdown bubbles begins. The primary bubble shifts downwards while

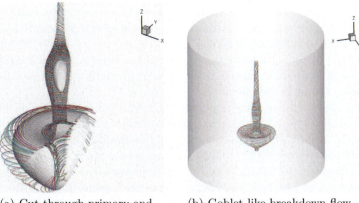

(a) Cut through primary and secondary breakdown bubble, Reynolds number 2240

(b) Goblet like breakdown flow structure, Reynolds number 2600

Fig. 3. Development of breakdown bubbles, visualized by streamlines and isosurfaces of the negative axial velocity component

separating from the secondary bubble, figure 3(a) and it reshapes to an inverted mushroom.

Although the second bubble rapidly shrinks, the streamlines are not so constricted as in the beginning of the whole simulation. This means that now fluid of the first bubble exhausts through the streamtube and prevents a further shrinking of the streamtube radius. The collapse of the secondary bubble is the next significant structure changing. In figure 3(b) the secondary breakdown bubble is completely vanished and the primary bubble becomes a mushroom like shape. Later on a vortex ring forms, which continuously becomes thinner until it finally disappears.

4 Topological Analysis

In this consideration topological flow analysis concentrates on the detection of stagnation points in the vector fields and on the calculation of their associated separatrices. Following Chong [7] and Dallmann [8] critical points are locations in the volume, where the vector, e.g. the velocity, vanishes and the vector field becomes singular. The appearance of these points is an indisputable sign that the structure of the flow has changed. Critical points cannot evolve alone: Usually they are born as twins, only in degenerated cases more than two new points can occur. The points are classified by their effect on the vector field in their vicinity. In principle in the case of a two dimensional vector field five different basic types may exist: node, saddle, focus, center and line. Often the last one is considered as a sequence of critical points. In three dimensions (3D) the type of a critical point is a mixing of three of the five basic types. The classification can be done by analysing the eigenvalues or invariants of the appendant tensor field. An extensive description of classification algorithms is given by Reyn [24] and Bakker et al. [1, 2]. Chong et al. [7], Perry et al. [23] used it in fluid mechanics applications to describe vortical structures. Dallmann [8, 9] used the critical point concept to identify flow separation and vortical structures in flow fields.

In this research the critical point detection algorithm works on tetrahedral grids, therefore, the hexahedral and prismatic grid cells were divided in tetrahedra. Then to find the critical points in the flow field a simple solver for linear equations can be applied for the corner point velocity vectors of each tetrahedron. As a next step the local separatrices of each detected critical point have to be integrated. Separatrices are integral lines of the current vector field with starting points nearby the critical point. "Nearby" means that the start points of integration are shifted a bit from the critical point, where the current vector entity vanishes, in direction of the "real" eigenvectors of the tensor of the appendant gradient field. At those start points the field is not singular anymore and a line integration can start. The principles of such an algorithm and its implementation can be found in the work of Helman et. al. [16] or Hesselink et al. [17]. Kenwright et al. [19] presented technical flow applications. An overview of topological fluid mechanics related work can be found in [29]. Newer developments in visualisation of critical points and flow structure are demonstrated by Sadarjoen [26] using sophisticated icon concepts. A combined approach of volume rendering technique and classical topological flow analysis of the vortex breakdown using critical points has been presented by Tricoche and Garth [30].

4.1 The Topology of Velocity

The topological analysis of the flow inside the cylinder starts with the velocity field. The underlying idea is to reduce the flow field information significantly by representing only critical points and their associated separatrices. Even though the flow structure is strongly reduced to only a few flow patterns, the main information is conserved. A first impression is given in figure 4(a), which shows the detected critical points (red coloured points) and their separatrices for the flow at a Reynolds number of 1600. As seen before the primary vortex breakdown bubble is fully developed, building a nearly perfect sphere. This observation will be used in the later analysis.

It is remarkable that the separatrices are not building a closed vortex bubble, in fact by passing the upper critical point they reveal the growing process of the breakdown bubble. Above the upper critical point the separatrices are encapsulating a small cylinder. This is the region, where the secondary bubble will occur later. The newly detected singular points validate that, which can be seen in figure 4(b).

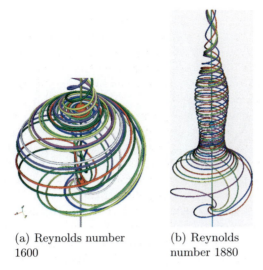

(a) Reynolds number 1600   (b) Reynolds number 1880

Fig. 4. Visualization of Vortex breakdown bubbles by critical points and their separatrices

4.2 Vortex Lines

The analysis of the velocity field demonstrates that considering the derivative fields of the velocity gives significant hints of the overall behaviour of the flow structure. Hence, the vorticity field has to be analysed. In contrast to the velocity field it is not possible to find isolated critical points in the vorticity field. Alternatively vortex line integration has been started to illustrate the structure of the vorticity field.

Looking closely to figure 5(a) the torsion of the vortex lines shows that below the vortexbreakdown bubble, they are weakly rotating with the same sense of rotation revolution like the streamlines in figure 1. Just before they reach the breakdown region their winding sense has been inverted in comparison to the winding of the streamlines, and their torsion is rapidly increased. After the vortex lines have left the region of the breakdown bubbles they are stretched and almost run parallel.

The same behaviour is depicted in figure 5(b). The reversed torsion can be well seen at the secondary bubble. Before they reach it they begin to loop in direction of the streamlines. Then they are skipping again in the other direction. After passing the bubble the vortex lines stretch again. Thus, a reversed (compared to the winding of the streamlines) winding of the vortex lines is connected with the vortex breakdown which is a generalisation of the theory of Darmofal [10]. He concentrated on the azimuthal vorticity component as a major parameter for vortex breakdown. Also Lopez et al. [21] showed the importance of the azimuthal vorticity component in regard to vortex breakdown conditions. The physical cause of this observation is the focal point of the following considerations.

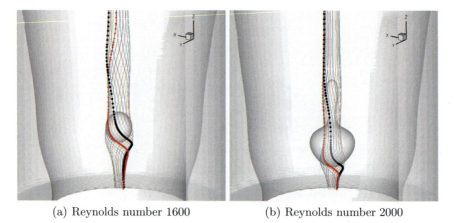

(a) Reynolds number 1600   (b) Reynolds number 2000

Fig. 5. Winded vortex lines surrounding the breakdown bubbles, stretched in the upper field

5 Local Flow Analysis

One aspect of the flow field calculation was to look at the temporal development of the main flow structures, which has been illustrated. Henceforth, the further analysis concentrates on the solution of the flow field at a certain time step: The flow state at a Reynolds number of 1600 is analyzed, when the first vortex breakdown bubble has been developed in a manner that two critical points can be found and their separatrices are forming a sphere like shape. Furthermore, we assume that the time dependent parts of the solution can be neglected in regard to the convective and diffusive parts, because we have chosen a very small time step in respect to the time scale made up by the acceleration of the lid and the occurring maximal velocity magnitude. Besides that we assume an axial symmetric flow. This is justified by the visualized results, we have shown above. Even when the vortex breakdown phenomenon occurs, the axisymmetry is well fulfilled, which could impressively be shown in the figures 1 and 4.

5.1 Streamfunction and Circulation near the Bubble

In order to analyse the flow locally near a vortex breakdown bubble, we use a system of cylindrical coordinates (r, φ, z) where the lower stagnation point is chosen as origin and the z - axis coincides with the vortex axis. We make use of the known fact that the radial and the axial velocity component of an incompressible axisymmetric flow can be derived by a streamfunction $\Psi(r, z)$ according to

$$v_r(r,z) = -\frac{1}{r}\frac{\partial \Psi(r,z)}{\partial z}, \qquad (2)$$

$$v_z(r,z) = \frac{1}{r}\frac{\partial \Psi(r,z)}{\partial r}. \qquad (3)$$

Note that streamlines are lying in the stream tubes $\Psi(r,z) = const.$

In addition to $\Psi(r,z)$ we consider the azimuthal velocity component $v_\varphi(r,z)$ as primary kinematic variable or, more suitably, the corresponding scalar field $\Gamma(r,z) := r \cdot v_\varphi(r,z)$ which we call circulation. Considering the relations between the velocity and the vorticity components in the case of an axisymmetric flow, it becomes obvious that $\Gamma(r,z)$ is related to the vorticity field $\omega := curl\mathbf{v}$ in the same way as the streamfunction to the velocity field:

$$\omega_r(r,z) = -\frac{1}{r}\frac{\partial \Gamma(r,z)}{\partial z}, \qquad (4)$$

$$\omega_z(r,z) = \frac{1}{r}\frac{\partial \Gamma(r,z)}{\partial r}. \qquad (5)$$

Thus, any surface $\Gamma(r,z) = const.$ is made up by vortex lines and represents a vortex tube.

Finally, the azimuthal vorticity component $\omega_\varphi(r,z)$ of an axisymmetric incompressible flow is connected with the streamfunction according to

$$\omega_\varphi(r,z) = -\frac{\partial}{\partial r}\left(\frac{1}{r}\frac{\partial \Psi(r,z)}{\partial r}\right) - \frac{1}{r}\frac{\partial^2 \Psi(r,z)}{\partial z^2}. \qquad (6)$$

Equation 6 is the axisymmetric counterpart of the well-known Poisson equation which connects the streamfunction to the vorticity in the case of a plane flow. It is worth mentioning that the kinematics relations 3–6 are valid for incompressible axisymmetric flows regardless of the material properties of the liquid. Indeed, they turn out to be useful also when analysing axisymmetric non-Newtonian flow fields (Böhme [3]). In the following we derive some new theoretical findings which allow us a deeper insight into the vector fields near the bubble and their interactions. We have to emphasize that the visualisation we have done before was the initial cause of the development of the following analysis. We are able to show on the basis of the Navier-Stokes equations that near a free axis both $\Psi(r,z)$ and $\Gamma(r,z)$ are even functions concerning r. This leads us to the following power series expansions near the lower stagnation point:

$$\Psi(r,z) = r^2(az + cr^2 + bz^2) + O(\varepsilon^5), \qquad (7)$$

$$\Gamma(r,z) = r^2(d + ez) + O(\varepsilon^4) \qquad (8)$$

where

$$\varepsilon = \sqrt{r^2 + z^2}. \qquad (9)$$

Making use of eqs. 2–5 we get the velocity components:

$$\begin{aligned} v_r(r,z) &= -r(a + 2bz) + O(\varepsilon^3), \\ v_\varphi(r,z) &= r(d + ez) + O(\varepsilon^3), \\ v_z(r,z) &= 2az + 2bz^2 + 4cr^2 + O(\varepsilon^3) \end{aligned} \qquad (10)$$

and the vorticity components

$$\begin{aligned}\omega_r(r,z) &= \quad -er + O(\varepsilon^2),\\ \omega_\varphi(r,z) &= -(2b+8c)r + O(\varepsilon^2),\\ \omega_z(r,z) &= \quad 2d + 2ez + O(\varepsilon^2)\end{aligned} \quad (11)$$

with a few independent coefficients only. Note that in contrast to the velocity the vorticity does not vanish at the critical point, but possesses an absolute summand in its z-component.

5.2 Identification of Expansion Coefficients

The local representation of the flow near the critical point given above consists of five terms and, thus, can be understood as the superposition of five elementary contributions: The terms with coefficients a, c, d and e represent an uniaxial elongational flow, a Poiseuille flow, a rigid body rotation and a torsional flow, respectively. The remaining terms connected with b may be interpreted as an axisymmetric flow towards a disk (at position $z = 0$) where the fluid adheres in accordance with a no-slip condition. The mentioned coefficients a, c, d and e rate the intensities of the different flow contributions.

We get some more insight into the kinematics restricting to situations where the vortex breakdown bubble is nearly spherical as in figure 4(a). Under this assumption the streamfunction $\Psi(r,z)$ vanishes on a sphere of radius r_0, see figure 6(a), which allows us to write the bracket within eq. 7 in the following form:

$$az + bz^2 + cr^2 = b((z-r_0)^2 + r^2 - r_0^2). \quad (12)$$

Thus, $c = b$ and $a = -2br_0$ which reduces the number of independent coefficients. Concerning the sign of the coefficients we bear in mind that the elongational rate

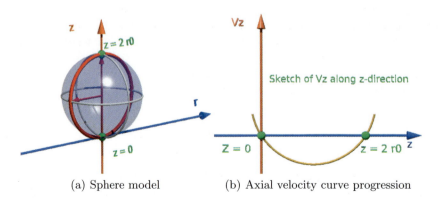

(a) Sphere model   (b) Axial velocity curve progression

Fig. 6. (a) Sketch of a breakdown bubble approximated as a sphere. (b) Sketch of the axial velocity component along the axis

$$2a = \left(\frac{\partial v_z(0,z)}{\partial z}\right)_{z=0} \tag{13}$$

is negative, if the flow approaches the critical point from below, $a < 0$. Hence, $b = c > 0$ in accordance with figure 6(b) as the coefficient 4b may be interpreted as the second derivative of the velocity along the axis:

$$4b = \left(\frac{\partial^2 v_z(0,z)}{\partial z^2}\right)_{z=0}. \tag{14}$$

Furthermore, the axial spin at the critical point,

$$d = \frac{1}{2}\omega_z(0,0), \tag{15}$$

is positive due to the assumed sense of rotation of the lid, $d > 0$, whereas the coefficient of the torsional flow contribution

$$e = \frac{1}{2}\left(\frac{\partial \omega_z(0,z)}{\partial z}\right)_{z=0} \tag{16}$$

is negative, $e < 0$, as the cross-section area of a vortex tube increases near the critical point, see figure 5.

5.3 The Dynamics of the Breakdown Bubble

After having introduced the approximate description of the flow kinematics in the breakdown bubble region we consider the corresponding information of the equation of motion

$$\frac{1}{2}grad\,|\mathbf{v}|^2 + \boldsymbol{\omega} \times \mathbf{v} = -\frac{1}{\rho}grad\,p - \nu curl\,\boldsymbol{\omega}. \tag{17}$$

ρ denotes the constant density of the fluid, $p(r,z)$ is the pressure field. Near the critical point where the velocity vanishes, the acceleration parts of eq. 17, $\frac{1}{2}grad\,|\mathbf{v}|^2$ and $\boldsymbol{\omega} \times \mathbf{v}$, do not have absolute summands, they are $O(\varepsilon)$ and therefore not relevant.

In contrast to that the specific viscous force has one absolute component, more precisely the specific viscous force is

$$-\nu curl\,\boldsymbol{\omega} = \begin{pmatrix} 0 \\ 0 \\ 4\nu(b+4c) \end{pmatrix} + O(\varepsilon). \tag{18}$$

Therefore, in the critical point the specific viscous force acts only in axial direction. Furthermore, on the vortex axis the pressure gradient also reduces to a z - component, reflecting the axial symmetry of the flow, which requires $\left(\frac{\partial p}{\partial r}\right)_{r=0} = 0$ in addition to $\frac{\partial p}{\partial \varphi} = 0$. Consequently, it has to be emphasized that at the critical point the axial pressure gradient is proportional to the viscous terms, given in eq. 18. Considering

$$\frac{1}{\rho}\left(\frac{\partial p(0,z)}{\partial z}\right)_{z=0} = 4\nu(b+4c), \qquad (19)$$

the positive coefficients b and c lead to a positive axial pressure gradient. By replacing b and c with the aid of eq. 14, we find the following remarkable relation:

$$\left(\frac{\partial p(0,z)}{\partial z}\right)_{z=0} = 5\rho\nu\left(\frac{\partial^2 v_z(0,z)}{\partial z^2}\right)_{z=0}. \qquad (20)$$

Thus the pressure gradient is simply connected with the second derivative of the axial velocity, i.e. with the curvature of the graph shown in fig. 6(b). In regard to spheres of different size it can be stated that for a small sphere like bubble the pressure gradient has to be higher than for large bubbles. This can be interpreted in a way that in case of initiation of a breakdown bubble a higher axial pressure gradient is needed than in the case of a growing bubble.

5.4 The Sign of the Circumferential Vorticity Component near the Axis

As mentioned above, we found that the vortex lines may change their winding sense near a vortex breakdown region, see figures 5. Hence, the azimuthal vorticity component ω_φ changes its sign along those vortex lines. In order to detect the conditions under which the phenomenon occurs, we consider the quantity $\frac{\omega_\varphi}{r}$ along the axis, i.e. in the limit $r \to 0$ at arbitrary position z, and link it to the axial gradient of the total pressure.

Starting with the expansion of the velocity component $v_z(r,z)$ near the axis,

$$v_z(r,z) = v_z(0,z) + \frac{1}{2}r^2\left(\frac{\partial^2 v_z(r,z)}{\partial r^2}\right)_{r=0} + O(r^4), \qquad (21)$$

we come to the following representation of the streamfunction near the axis of an axisymmetric flow:

$$\Psi(r,z) = \frac{1}{2}r^2 v_z(0,z) + \frac{1}{8}r^4\left(\frac{\partial^2 v_z(r,z)}{\partial r^2}\right)_{r=0} + O(r^6). \qquad (22)$$

Inserting this into eq. 6 we get

$$\lim_{r \to 0} \frac{\omega_\varphi(r,z)}{r} = -\left(\frac{\partial^2 v_z(r,z)}{\partial r^2}\right)_{r=0} + \frac{1}{2}\frac{\partial^2 v_z(0,z)}{\partial z^2} \qquad (23)$$

$$= -\frac{1}{2}\left(\Delta v_z(r,z)\right)_{r=0}$$

where Δ indicates the Laplace operator. The last equality follows in connection with eq. 21 which states that we may substitute

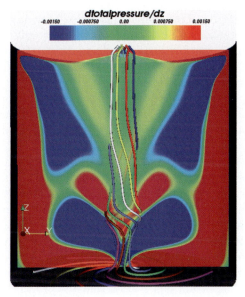

Fig. 7. Axial gradient of total pressure on a meridional cut, additionally vortex lines visualised, Reynolds number 1880

$$\left(\frac{1}{r}\frac{\partial v_z(r,z)}{\partial r}\right)_{r=0} = \left(\frac{\partial^2 v_z(r,z)}{\partial r^2}\right)_{r=0} . \qquad (24)$$

The equation of motion 17 allows us to eliminate the right side of eq. 23:

$$\nu(\Delta v_z(r,z))_{r=0} = \frac{\partial}{\partial z}\left(\frac{p(0,z)}{\rho} + \frac{1}{2}v_z^2(0,z)\right) . \qquad (25)$$

Thus, we find that the following striking relation is valid along the axis of any axisymmetric, incompressible, viscous flow:

$$\lim_{r \to 0} \frac{\omega_\varphi(r,z)}{r} = -\frac{1}{2\rho\nu}\frac{\partial}{\partial z}\left(p(0,z) + \frac{\rho}{2}v_z^2(0,z)\right) . \qquad (26)$$

Among other things it tells us that vortex lines near the axis change their winding sense at those positions where the axial gradient of the total pressure $p+\frac{\rho}{2}v_z^2$ changes its sign. This indeed may happen several times along the axis between the bottom of the cylinder and the rotating lid, see figure 7.

6 Conclusion

Bearing the overall goal in mind to find mechanisms which allow to control the vortical flow either in order to prevent vortex breakdown or to generate it, we analysed the vortex breakdown phenomenon occurring in a cylinder with an instationary rotating lid. This simple cylindrical configuration allowed us to study the vortex breakdown phenomenon depending only on one parameter, the Reynolds number by fixing the height-to-diameter ratio of one. In the numerical simulation the angular velocity of the lid was gradually increased: In the beginning no vortex breakdown bubble occurred, while exceeding a certain Reynolds number one vortex breakdown bubble occurs, later on a second one. In the course of further increasing the angular velocity the secondary bubbles disappears, then also the primary bubble and at last the complete breakdown structure vanishes.

A topological flow analysis was done for the velocity field and extended to the vorticity field. A comparison of the streamlines and vortex lines shows that the vortex breakdown bubbles are correlating to a reversed torsion of the vortex lines. Reversed means that the sense of revolution of the vortex lines is contrary to that of the streamlines. Furthermore, based on an appropriate visualization it could be shown that the winding of the vortex lines have changed their revolution just before the breakdown begins. Only this observation has led to a new starting point for further analysis and discussion. Together with the important observation that the primary vortex breakdown bubble has approximately the shape of a sphere this has provided a new analytical description of the phenomenon. Kinematic considerations concerning the streamfunction led to a deduction of the dynamics of the breakdown bubble at the critical point. It could be shown that in the singular points viscous forces lead to a positive axial pressure gradient. Furthermore, it could be visualized that vortex lines change their winding sense at those positions where the axial gradient of the total pressure changes its sign. Again we have to emphasize that without a proper comparative visualization of the breakdown bubbles, of streamlines and vortex lines we would not have been got the essential ideas for our flow field analysis.

Further work is needed in order to develop a formal theory, which explains and quantifies the topological and chronological interaction between the different vector fields. Therefore, additional experiments, computational simulations and advanced visualization techniques are required for developing a complete theory of vortex breakdown. This work, so far, has focused on the behaviour of vortex breakdown bubbles in the confined container flow, may be, the results will initiate further studies on the vortex breakdown mechanism of free unbounded vortices.

References

1. Bakker, P. G.: Bifurcations in Flow Patterns. PhD thesis, Delft University of Technology, 1988
2. Bakker, P. G., de Winkel, M. E. M.: On the topology of three-dimensional separated flow structures an local solutions of the Navier-Stokes equations. Topological Fluid Mechanics, Proc. of the IUTAM Symposium, Cambridge, Aug. 13–18, 1989

3. Böhme, G.: Strömungsmechanik nichtnewtonscher Fluide, B.G. Teubner, Stuttgart 2000
4. Böhme, G., Rubart, L., Stenger, M.: Vortex breakdown in shear-thinning liquids: experiment and numerical simulation. J. Non-Newt. Fluid Mech. 45 (1992), pp. 1–20
5. Brons, M., Voigt, L. K., Sorensen, J. N.: Streamline topology of steady axisymmetric vortex breakdown in a cylinder with co- and counter-rotating end-covers. J. Fluid Mech. (1999), vol. 401, pp. 275–292
6. Chabral, B.; Leedom, L. C.: Imaging Vector Fields Using Line Integral Convolution. In: Proceedings of SIGGRAPH '93. New York, 1993, pp. 263–270
7. Chong, M. S., Perry, A. E., Cantwell, B. J.: A general classification of three-dimensional flow fields. Phys. Fluids A, vol. 2, no. 5, 1990, pp. 261–265
8. Dallmann, U.: Three-dimensional vortex structures and vorticity topology. Proceedings of the IUTAM Symposium on Fundamental Aspects of Vortex Motion, Tokyo, Japan, 1987
9. Dallmann, U.: Analysis of Simulations of Topologically Changing Three-Dimensional Separated Flows. IUTAM Symposium on "Separated Flows and Jets", Novosibirsk, 1990
10. Darmofal, D. L.: The Role of Vorticity Dynamics in Vortex Breakdown, AIAA 93-3036
11. Delery, J. M.: Aspects Of Vortex Breakdown. Prog. Aerospace Sci, vol. 30, pp. 1–59, 1994
12. Escudier, M. P.: Observations of the flow produced in a cylindrical container by a rotating endwall. Exps. Fluids 2 (1984), pp. 189–196
13. Escudier, M. P.: Vortex Breakdown: Observation And Explanation. Prog. Aerospace Sci Vol. 25, pp. 189–229, 1988
14. Faler, J. H., Leibovich, S.; An experimental map of the internal structure of a vortex breakdown. J. Fluid Mech. (1978), vol 86, pp. 313–335
15. Goldshtik, M., Hussain, F.: The nature of vortex breakdown, Phys. Fluids 9, pp. 263–265, 1997
16. Helman, J., Hesselink L.: Visualizing vector field topology in fluid flows. IEEE Computer Graphics and Applications, vol. 11, no. 3, May 1991, pp. 36–46
17. Hesselink, L. Delmarcell, T.: Visualization of vector and tensor data sets, In Scientific visualization - Advances and challenges, L. Rosenblum, R. A. Earnshaw, J. Encarnacao, H. Hage, A. Kaufman, S. Klimenko, G. Nielson, F. Post and D. Thalman, Eds. Academic Press, 1994, pp. 367–390
18. Hussain, H. S., Hussain, F., Goldshtik, M.: Anomalous separation of homogeneous particle-fluid mixture: Further observations. Phys. Rev. E 52, pp. 4909–4923, 1995
19. Kenwright, D. N., Henze, C., Levit, C.: Feature Extraction of Separation and Attachment Lines. IEEE Transactions on Visualization and Computer Graphics, vol. 5, no. 2, 1999
20. Leibovich, S.: The structure of vortex breakdown, Ann. Rev. Fluid Mech. 10, pp. 221–246, 1978
21. Lopez J. M., Brown G. L.: Physical Mechanism Of Axisymmetric Vortex Breakdown. Tenth Australasian Fluid Mechanics Conference, University of Melbourne, 1989
22. Lugt, H. J., Hausling, H. J.: Axisymmetric vortex breakdown in rotating fluid within a container, ASME J. Appl. Mech. 49(4), pp. 921–922, 1982

23. Perry, A. E., Chong, M. S.: A description of eddying motions and flow patterns using critical-point concepts. Ann. Rev. Fluid Mech., vol. 19, 1987, pp. 125–155
24. Reyn, W. J.: Classification and Description of the Singular Points of a System of Three Linear Differential Equations. ZAMP, vol. 15, 1964
25. Sarpkaya, T.: Vortex Breakdown and Turbulence, AIAA 95–0433
26. Sardajoen, I. A.: Extraction and Visualization of Geometries in Fluid Flow Fields. PhD thesis, Delft University of Technology, 1999
27. Sotiropoulos, F., Ventikos, Y.: The three-dimensional structure of confined swirling flows with vortex breakdown. J. Fluid Mech. (2001), vol. 426, pp. 155–175
28. Spohn, A., Mory, M., Hopfinger, E.J.: Experiments on vortex breakdown in a confined flow generated with rotating bottom disk. J. Fluid Mech. (1998), vol. 370, pp. 73–99
29. Theisel, H., Weinkauf, T., Hege H.-C., Seidel, H.-P.: Stream Line and Path Line Orientated Topology for 2D Time-Dependent Vector Fields, In *Proceeding IEEE Visualization 2004*, pp. 321–328, 2004
30. Tricoche X., Garth C., Kindlmann G., Deines E., Scheuermann G., Rütten M., Hansen C.: Visualization of Intricate Flow Structures for Vortex Breakdown Analysis. In *Proc. of the IEEE Visualization '04 Conf.*, pp. 187–194, 2004

Flow Topology Beyond Skeletons: Visualization of Features in Recirculating Flow

Ronald Peikert and Filip Sadlo

Computer Graphics Laboratory, Computer Science Department,
ETH Zurich, Switzerland, {peikert, sadlo}@inf.ethz.ch

Summary. A pattern often found in regions of recirculating flow is the vortex ring. Smoke rings and vortex breakdown bubbles are two familiar instances of this pattern. A vortex ring requires at least two critical points, and in fact this minimum number is observed in many synthetic or real-world examples. Based on this observation, we propose a visualization technique utilizing a Poincaré section that contains the pair of critical points. The Poincaré section by itself can be taken as a visualization of the vortex ring, especially if streamlines are seeded on the stable and unstable manifolds of the critical points. The resulting image reveals the extent of the structure, and more interestingly, regions of chaos and islands of stability. As a next step, we describe for the case of incompressible flow an algorithm for finding invariant tori in an island of stability. The basic idea is to find invariant closed curves in the Poincaré plane, which are then taken as seed curves for stream surfaces. For visualization the two extremes of the set of nested tori are computed. This is on the inner side the periodic orbit toward which the tori converge, and on the outer side, a torus which marks the boundary between ordered and chaotic flow, a distinction which is of importance for the mixing properties of the flow. For the purpose of testing, we developed a simple analytical model of a perturbed vortex ring based on Hill's spherical vortex. Finally, we applied the proposed visualization methods to this synthetic vector field and to two hydromechanical simulation results.

1 Introduction

Vector field topology, introduced by Helman and Hesselink [7], can be summarized as the use of concepts from the theory of continuous dynamical systems (see e.g. [3]) in scientific visualization. The main motivation for vector field topology is its ability to provide a condensed representation of a vector field. The most popular such representation is the topological skeleton which is usually defined as the set of all critical points and all separatrices. In two dimensions, the topological skeleton provides a segmentation of the domain into regions of similar flow behavior. The separatrices can be obtained by computing the stable and unstable manifolds of all critical points of saddle type. However, unless the vector field is irrotational, there may also exist periodic orbits that behave like sources or sinks. If this is the case, the topological skeleton computed this way is incomplete. Only if the set of (isolated) periodic orbits is explicitly added to the skeleton, the full segmentation is obtained. An algorithm for finding isolated periodic orbits was developed by Wischgoll and Scheuermann [25].

When going to three dimensions, the topological skeleton can again be defined as the set of all critical points and all separatrices. The separatrices are the stable and unstable manifolds of saddles and spiral saddles (saddle foci), coming in pairs of a 1D and a 2D manifold, i.e. a streamline and a stream surface. The 1D manifolds are obviously not very useful for the purpose of segmenting a 3D domain. Only in the case of spiral saddles, they have some relevance, as they are sometimes understood as vortex core lines. The 2D manifolds theoretically provide segmentation, but in practical flows, these stream surfaces can become very convoluted. An alternative is to show only their pairwise intersections, known as saddle connectors [23] or heteroclinic orbits, resulting in a visualization of the connectivity between critical points.

The usage of vector field topology for scientific visualization is not restricted to showing topological skeletons. For example, critical points can be used for streamline placement [26]. Even if the full set of critical points is used without any type analysis, this strategy was shown to yield effective visualizations by Weinkauf et al. [24]. Alternatively, a visualization of the local flow behavior near critical points can be obtained by displaying icons showing the linearized flow defined by the critical point type and by the eigenvectors of the Jacobian of the vector field [4]. The same information can be used to seed short streamlines near critical points [9], giving a slightly more global picture of the flow.

It is interesting to notice that most work done so far in topology-based visualization falls in one of two categories, either giving a global picture of the entire domain or a local picture of neighborhoods of critical points. While global effects are an interesting part of dynamical systems and chaos theory, it can be argued that for flow visualization, they are less relevant because of issues such as domain boundaries, simulation accuracy, or time-dependence. But also the other extreme, independent visualization of critical points, can be regarded as unsatisfactory, since much of the topological information is left unused. We believe that vector field topology has much to offer for flow structures which fall in between the two extremes. One such structure is the vortex ring, which is essentially determined by two critical points and a small number of periodic orbits. In an earlier paper [17], we used a specialized stream surface algorithm for the visualization of such middle-scale flow features. Garth et al. [5] and Tricoche et al. [20] demonstrated how complex flow structures such as vortex breakdown bubbles can effectively be visualized by using stream surfaces and volume rendering, respectively. In this paper we present complementary visualization techniques which are more closely oriented at the topology.

There has of course been previous work on visualization of dynamical systems. In particular for visualizing the behavior near critical points of a 3D system, Löffelmann et al. introduced various techniques such as glyphs [9], Poincaré maps [11] directly visualized in the context of the 3D field, and bundles of trajectories [10] rendered as illuminated streamlines [19]. In all these cases, the object to be visualized was a given dynamical system. What we show in this work is that vector fields originating from other sources, such as synthetic flow fields or industrial CFD results, are just as well suited for being visualized as dynamical systems. In particular, we believe that is worth looking at further concepts of the dynamical systems theory than those which have made their way into the toolbox of vector field topology. As a source of inspiration, the book by Abraham and Shaw [1] can be recommended.

2 Topology of vortex rings

A typical feature occurring in recirculation regions is a connected pair (C_0, C_1) of critical points where C_0 is a 1:2 spiral saddle (1 incoming and 2 outgoing dimensions) and C_1 is a 2:1 spiral saddle. By "connected" we mean that the 2D unstable manifold $W^u(C_0)$ and the the 2D stable manifold $W^s(C_1)$ intersect. The intersection is then a set of saddle connectors. If the spiraling at both C_0 and C_1 is sufficiently strong, the surface pair $(W^u(C_0), W^s(C_1))$ roughly delimits a recirculation region. In its simplest form this region is a vortex ring, as is illustrated in Figure 1. The saddle connectors alone give already some idea of the geometry of the recirculation region. However, there is usually more topological information available for visualization than just the saddle connectors. Such features include chaotic regions, islands of stability, and invariant tori having rational or irrational rotation numbers (i.e. frequency ratios).

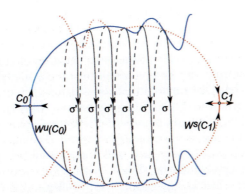

Fig. 1. Unstable manifold (solid) and stable manifold (dotted) of spiral saddles C_0 and C_1, respectively. Their intersection is a pair of saddle connectors σ and σ'

If the (3D) vector field is divergence-free any such transversal intersection of the 2D (un-)stable manifolds of two spiral saddles with sufficient spiraling automatically implies a heteroclinic tangle. This phenomenon which is also known as Shilnikov chaos [15, 18] is well known in the dynamical systems literature and can be described as follows. In general, the two manifolds $W^u(C_0)$ and $W^s(C_1)$ do not coincide, but intersect transversally. In this case they intersect at an even number of saddle connectors, usually a pair σ and σ' of them. Between the windings of the saddle connectors, the manifolds form two "tubes" that are wrapped around the structure. The tubes have constant flux (i.e. independent of cross sections) because the 2D manifolds are stream surfaces, and the sum of the two fluxes is zero because of the divergence-free condition. This implies that toward the critical points, where velocities approach zero, the tubes must either have increasing cross section area or develop folds that extend into regions of higher velocities. These folds, known as lobes, are typical of vortex breakdown bubbles (see e.g. [16]). It might seem strange to use the term vortex ring not only for structures such as smoke rings but also for the chaotic structure of a vortex breakdown bubble. However, this is consistent with the literature [8].

Much of dynamical systems theory deals with the special case of Hamiltonian systems, because of their area-conserving maps which are mainly responsible for chaotic behavior. Among the vector fields, the divergence-free ones play a similar role, and in fact they are related to Hamiltonian systems. In 2D, divergence free-vector fields (written as ODEs) and Hamiltonian systems are even the same, with the stream function Ψ (with $\frac{\partial \Psi}{\partial x} = -\dot{y}$ and $\frac{\partial \Psi}{\partial y} = \dot{x}$) playing the role of the Hamiltonian function. In 3D, a divergence-free vector field is volume preserving, but does not necessarily have area-conserving Poincaré maps. Nevertheless, the Poincaré map is at least flux-conserving, which is the reason for the above mentioned Shilnikov chaos to occur.

The use of topological methods for time-dependent flow is sometimes questioned. Haller [6] says that structures such as chaotic tangles or KAM tori (i.e. invariant tori of a Hamiltonian system) do not exist in finite-time turbulent data sets. Nevertheless we believe that it is interesting to search for such structures, first of all in steady flow fields (where time can be viewed as infinite). It can be demonstrated that these topological features exist in practical flow data, meaning that the catalog of features to be studied in vector field topology must include invariant tori, chaotic regions, intersecting stable and unstable manifolds and multiple saddle connectors. Clearly, the definition of stable and unstable manifolds requires infinite-time flows, but this already holds for the separatrices in the commonly treated 2D case. Practical flow has often small enough time-dependence that their visualization as steady flow is a good enough approximation. The fact that vortex breakdown bubbles have been photographed in experiments [16] confirms that this holds even if chaos is involved. Furthermore, the shapes observed in experiments have been shown to be consistent with the manifolds of critical points in a steady vector field [18].

3 Analytical vortex ring model

For testing our algorithms, we developed a simple analytic vortex ring model based on Hill's spherical vortex (see e.g. [14]). An analytical vector field has the advantage that artifacts due to discretization and interpolation can be excluded. A second motivation was to demonstrate that a rich topology (Figure 7) is possible even if the vector field has only two critical points and can be expressed with only quadratic terms (Eq. 3).

An instance of Hill's spherical vortex can be described by the two velocity fields

$$\mathbf{u}^i(x, y, z) = \begin{pmatrix} xz \\ yz \\ z^2 + 1 - 2r^2 \end{pmatrix} \quad (1)$$

for points inside the unit sphere $r = \sqrt{x^2 + y^2 + z^2} <= 1$ and

$$\mathbf{u}^o(x, y, z) = \begin{pmatrix} xzr^{-5} \\ yzr^{-5} \\ z^2 r^{-5} - \frac{1}{3}r^{-3} - \frac{2}{3} \end{pmatrix} \quad (2)$$

for points outside it ($r >= 1$).

The field is divergence-free, and it solves the Navier-Stokes equations (together with a matching pressure field). Furthermore, the field has zero vorticity outside the unit sphere. See Figure 2.

By adding a swirl $(wy, -wx, 0)$, a rotating vortex ring model is obtained. This simple model does no more solve the Navier-Stokes equations but is capable of generating the topological phenomena that can be observed in vortex rings. Physically correct variants of Hill's vortex with swirl exist, but they are more expensive to compute since Bessel functions have to be evaluated [14]. A different kind of generalization of Hill's spherical vortex are the Norbury

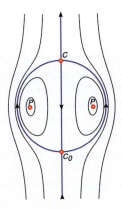

Fig. 2. Hill's spherical vortex (axial slice). C_0, C_1: critical points (spiral saddles), P: periodic orbit

Fig. 3. Hill's spherical vortex with swirl ($\omega = 2\pi$) and tilt ($\epsilon = 0.313$). Slice of the stable manifold of the critical point at $(0,0,1)$

vortex rings [13] where the vorticity is confined to toroidal regions instead of the sphere.

In order to obtain the chaotic behavior of a real vortex ring, the symmetry must be broken. In our model we do this by tilting the x-axis, which is motivated by experimental studies of vortex rings (see [21]). By substituting $z' = z + \epsilon x$ for z and $w' = w + \epsilon u$ for w in Eq. 1, and by adding the swirl, we get the velocity fields

$$\mathbf{u}^i_{\epsilon\omega}(x,y,z) = \begin{pmatrix} \omega y \\ -\omega x \\ 0 \end{pmatrix} + \begin{pmatrix} xz' \\ yz' \\ zz' + 1 - 2r'^2 \end{pmatrix} \quad (3)$$

for points inside the distorted unit sphere $r' = \sqrt{x^2 + y^2 + z'^2} <= 1$ and

$$\mathbf{u}^o_{\epsilon\omega}(x,y,z) = \begin{pmatrix} \omega y \\ -\omega x \\ 0 \end{pmatrix} + \begin{pmatrix} xz'r'^{-5} \\ yz'r'^{-5} \\ zz'r'^{-5} - \frac{1}{3}r'^{-3} - \frac{2}{3} \end{pmatrix} \quad (4)$$

for points outside of it.

This modified field is still divergence-free. It can be shown that the only critical points are two spiral saddles at $(0,0,-1)$ and $(0,0,1)$.

Figure 6 shows a $x = 0$ slice of the unstable manifold of the critical point at $(0,0,1)$, computed by seeding 200000 streamlines near the critical point and allowing for a maximum of 200000 intersections with the plane. The coloring of intersection points represents time, expressed in number of intersections with the plane. A rainbow color map is used, starting with violet and ending with red for intersection number 1000 and above. The system of three ODEs was solved with the 4th order Runge-Kutta-Fehlberg routine from the Netlib library.

If an even simpler model is needed, it is also possible to use just the inner part $\mathbf{u}^i_{\epsilon\omega}$ for the entire domain, see Figure 7.

4 Visualization techniques for vortex rings

The visualization technique we propose for vortex rings consists of three steps. First, the set of critical points is computed, and candidates for vortex rings are generated among pairs of spiral saddles of opposite type. Then, a plane passing through the two critical points is chosen, and a Poincaré section of $W^u(C_0)$ and $W^s(C_1)$ is taken. If an intersection of these is observed, the vortex ring is confirmed. Finally, the Poincaré section is used to extract islands of stability, i.e. to segment regions of chaotic and ordered flow.

4.1 Detection of vortex rings

The set of critical points is computed with the standard cell-by-cell method. Only cells where all three vector components have a zero crossing have to be processed. For classifying the critical points, the eigenvalues of the Jacobian are needed. One positive real eigenvalue and a pair of complex eigenvalues with negative real parts indicate a 2:1 spiral saddle, while opposite signs indicate a 1:2 spiral saddle. Pairs (C_0, C_1) of these two kinds of spiral saddles are now taken as candidates for vortex rings. We choose pairs simply based on vicinity and leave it to the verification step described in Section 4.2 to eliminate wrong pairs. Alternatively one could extract vortex core lines and make use of the fact that critical points of spiral saddle type lie on core lines because they fulfill both the Sujudi-Haimes and Levy criterion. Yet another approach would be to compute the set of saddle connectors which gives the correct pairs directly.

4.2 Poincaré section

We choose a plane passing through C_0 and C_1, using the remaining degree of freedom to fit the plane to the two real eigenvector directions of the two critical points. This way, the section is taken close to the center line of the vortex ring. Then a uniform grid is defined on the plane with an extent chosen based on the distance $d = ||C_1 - C_0||$. We found a square with edge length $2d$ to be sufficient in most cases. The two manifolds $W^u(C_0)$ and $W^s(C_1)$ are now computed based on a discrete set of seed points, and the intersections with the Poincaré section are stored as two (texture) images. Seed points for the manifold of, say, C_0 are generated as follows. A first seed s_0 is chosen at a small offset from C_0 on the Poincaré plane where it intersects the plane spanned by the two complex eigenvectors. From s_0, a streamline is integrated in the time direction where the distance from C_0 increases. Its next iterate (i.e. intersection with the Poincaré plane) is denoted by s_1. Further seed points are now generated on the straight line segment between s_0 and s_1 by logarithmically interpolating the distance of the seed points to C_0. Logarithmic interpolation is appropriate because close to C_0 streamlines are logarithmic spirals, and the error introduced by interpolating along a straight line falls off with the streamlines converging to the 2D manifold.

Integrating streamlines for all seed points and for a given maximum number of intersections with the Poincaré plane results in an image showing the intersection curve of the 2D manifold. By overlaying the images of C_0 and C_1 it can be decided if the manifolds intersect. An example pair is shown in Figure 10. In that case the image shows the lobes (folds) extending toward the second critical point. It also shows the chaotic region formed by the inward extending lobes, and it typically shows a hierarchy of islands of stability. The islands of stability are toroidal regions around a periodic orbit of minimal period. The inner part of stability islands is typically filled with nested invariant tori with no flux across them (stream surfaces, known as KAM tori in the case of Hamiltonian systems). Further out, chains of secondary islands can often be seen. These can be separated from the primary island by first regions of chaos. When the chaotic region is reached, so-called cantori [12] can appear. These are porous tori of measure zero, which in some cases (if the rotation number is a "noble" irrational number) have very little flux across them, and act therefore as partial barriers.

4.3 Islands of stability

From the previous step the Poincaré sections of $W^u(C_0)$ and $W^s(C_1)$ are now given as scalar fields on a regular 2D grid (or image) where the data values (or color indices) store the integration time or zero for cells that were not intersected. The goal is now to segment in the overlay of the two images the islands of stability. First, to clean the boundaries, a morphological closure operation is performed. This is followed by a component labeling step. Any component which does not extend to the image boundary is now checked for being an island of stability. A problem here is to distinguish islands of stability from holes that are formed by inward folding lobes. It can be observed that the latter are reached after much shorter integration time, hence when the average data value on their boundary is computed, this value is small compared to that of stability islands (see Figures 6, 7, 9, 10).

The obtained candidates for islands of stability are now processed in order of decreasing size. First, a streamline is seeded at the center of the island's bounding rectangle and whenever the Poincaré plane is intersected, the labeled component of the intersection point is marked as being part of the same island. If the streamline intersects the Poincaré section at a point outside of a component with a valid label, the test has failed.

Given now an island of stability, we want to visualize its internal structure which is a periodic orbit surrounded by nested invariant tori, with possible island chains interspersed in the outer part. For the Hill's vortex example, the primary and secondary islands are shown in Figure 4.

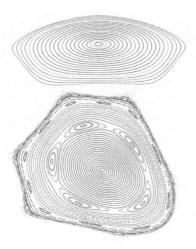

Fig. 4. Internal structure of primary and secondary island of Figure 3

We will visualize as two characteristic features the periodic orbit in the center and the outermost torus. The streamline seeded at the center of the island's bounding rectangle is integrated for a few "rounds" (detectable by increasing/decreasing x and y coordinates in the Poincaré plane). This should produce a set of points lying densely on a closed curve, otherwise it has to be retried from a slightly offset seed point. If a closed curve is obtained, the center of its bounding rectangle can be used for the next iteration of the process which is repeated until a fixed point is found.

This algorithm exploits the special structure of nested tori and is significantly faster than the general approach of looking for fixed points of the Poincaré map, especially since in the case of secondary islands no fixed points are found and successive powers of the Poincaré map must be computed and searched for fixed points, too.

For finding the boundary of the island of stability, an iterative search is started with a seed curve consisting of the outermost black (zero) pixels. At pixels which are mapped to a pixel outside the boundary, the seed curve is corrected inward by a pixel. This is repeated until all pixels of the seed curve are mapped to pixels inside the island. Finally, on these pixels the map is iterated a few times in order to reach a fixed curve. The obtained curve can be used as a seed curve for a simplified stream surface algorithm which requires only integration until the same component of the Poincaré plane is intersected again. Figure 5 shows a pair of stream surfaces obtained this way.

With a similar technique, the manifolds $W^u(C_0)$ and $W^s(C_1)$ can be obtained as stream surfaces with seed curves extracted from the Poincaré section. The stream surface can of course be computed directly, but this requires a robust algorithm to cope with the highly curved lobes.

Fig. 5. Primary (yellow) and secondary (red) islands rendered as stream surfaces

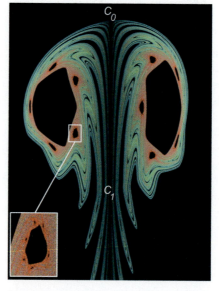

Fig. 6. Hill's spherical vortex with swirl ($\omega = 2\pi$) and tilt ($\epsilon = 0.313$). Slice of the stable manifold of the critical point at $(0, 0, 1)$

5 Results

We applied the techniques described in Section 4 to two CFD simulation results. In both cases, the data are given at the nodes of unstructured hexahedral grids. In principle, the computed velocity fields are divergence-free, however this is only true for the integrals over the control volumes, but not for the trilinearly interpolated data. Since we observed that any residual

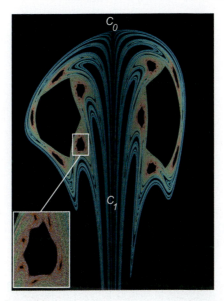

Fig. 7. Inner part $\mathbf{u}^i_{\varepsilon\omega}$ of Hill's spherical vortex with swirl ($\omega = 2\pi$) and tilt ($\epsilon = 0.442$)

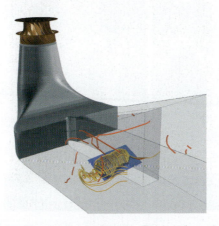

Fig. 8. Overview of the flow in the draft tube. Poincaré section used for Figure 10 shown as blue rectangle, vortex core lines shown in red

divergence left in the data causes the chaotic region to shrink, we did a divergence cleaning of the data prior to the visualization. The standard method for divergence cleaning is the Hodge projection method [2, 22] which is based on the decomposition of $\mathbf{u}$ into a divergence-free part and an irrotational part, $\mathbf{u} = \mathbf{u}_0 + \nabla s$. It follows $\nabla \cdot \mathbf{u} = \nabla \cdot \nabla s$ which is a Poisson equation for s.

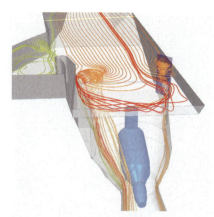

Fig. 9. Overview of the flow in the river power plant. Poincaré section used for Figure 11 shown as blue rectangle

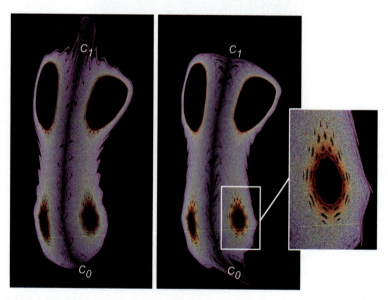

Fig. 10. Stable (left) and unstable (right) manifolds of vortex ring in draft tube dataset

5.1 Vortex ring in Francis draft tube

In the time-dependent simulation of the draft tube of a Francis turbine, we found a vortex ring extending spanwise and with a temporally quite stable behavior. An overview of the flow with the vortex ring and the rectangle used for the Poincaré section can be seen in Figure 8. The stable and unstable 2D manifolds of the two critical points show the structure of the vortex ring with

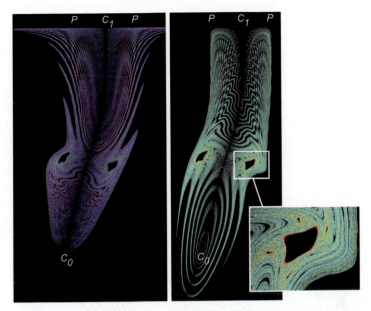

Fig. 11. Left: Stable manifold of spiral saddle C_0 in river power plant dataset. Right: unstable manifold of periodic orbit P, approximated by seeding just below spiral saddle C_1, close-up on primary island of stability

two primary islands of stability, see Figure 10. The abrupt change of colors near the islands of stability corresponds to jumps in integration time and therefore indicates cantori. These are toroidal surfaces which act as partial barriers for the mixing of the fluid.

In an earlier paper [17], we visualized the same flow structure with a volumetric technique but without divergence cleaning. As a result, most of the chaotic folding was lost because the flow was quickly attracted to a toroidal surface.

5.2 Vortex ring in simulation of a river power plant

Our second example is the flow in a river power plant developing two large vertical vortices at the surface, see Figure 9. We selected the left one of them, and chose a Poincaré section in the vertical plane through the two critical points. The result is shown in Figure 11.

In this example, the vortex ring extends to the (free slip) water surface where one of the two spiral saddles is located. The unstable manifold of the latter coincides with the stable manifold of a periodic orbit of saddle type which is also located at the water surface. In order to be able to integrate streamlines at the water surface, the normal velocity component had to be set to exactly zero, i.e. residual normal velocities from the simulation had to be removed.

The seemingly ring-shaped lobes are an artifact of the slice plane which does not follow well the curved center line of the structure. The effective shape of the lobes is similar to the one in Figure 4.

6 Conclusion

We presented an algorithm for finding vortex rings in velocity fields and visualizing them by means of a Poincaré section. Based on the latter, we described how islands of stability can be identified and seed curves for invariant tori are obtained, in particular for the outermost of the nested tori. A fast method was presented for computing the central periodic orbit of an island of stability. By applying these techniques to CFD data, we were able to find vortex rings and visualize them. Finally, we developed an analytical model of a perturbed vortex ring.

Part of the underlying theory requires divergence-free vector fields and thus incompressible flow. However, some of the proposed visualization techniques are also applicable to compressible flow. As an interesting future work we see the application of the proposed techniques, possibly modified, to examples of compressible flow such as smoke rings.

Although most of the vortex rings we found in CFD results contain just two critical points, some others have four or more of them. Additional critical points appear during events such as merging or splitting of vortex rings. Often there are small additional vortex rings which exist only for a short time and can thus be considered as noise. It would be an interesting topic to study how the various topology simplification techniques could improve our visualization technique.

Acknowledgment

We thank the anonymous reviewers for their valuable suggestions and France Suerich-Gulick for the river power plant data. This work was supported by the Swiss Commission for Technology and Innovation (CTI) under grant 7338.2 ESPP-ES.

References

1. Abraham R. H., Shaw C. D.: *Dynamics, the Geometry of Behavior*. 2nd ed. Addison-Wesley, 1992.
2. Brackbill J., Barnes D.: The effect of nonzero $\nabla \cdot \mathbf{B}$ on the numerical solution of the magnetohydrodynamic equations. *J. Comput. Phys. 35* (1980), 426–430.
3. Guckenheimer J., Holmes P.: *Nonlinear Oscillations, Dynamical Systems and Bifurcations of Vector Fields*. Applied Mathematical Sciences, Vol. 42. Springer, New York, Berlin, Heidelberg, Tokyo, 1983.

4. Globus A., Levit C., Lasinski T.: A tool for visualizing the topology of three-dimensional vector fields. In *Proc. IEEE Visualization '91* (1991), pp. 33–40.
5. Garth C., Tricoche X., Salzbrunn T., Bobach T., Scheuermann G.: Surface techniques for vortex visualization. In *VisSym* (2004), pp. 155–164, 346.
6. Haller G.: Lagrangian structures and the rate of strain in a partition of two-dimensional turbulence. *Phys. Fluids 13* (2001), 3365–3385.
7. Helman J., Hesselink L.: Representation and display of vector field topology in fluid flow data set. *IEEE Computer* (August 1989), 27–36.
8. Krasny R., Fritsche M.: The onset of chaos in vortex sheet flow. *J. Fluid Mech. 454* (2002), 47–69.
9. Löffelmann H., Doleisch H., Gröller E.: Visualizing dynamical systems near critical points. In *14th Spring Conference on Computer Graphics* (April 1998), Kalos L. S., (Ed.), pp. 175–184.
10. Löffelmann H., Gröller E.: Enhancing the visualization of characteristic structures in dynamical systems. In *Visualization in Scientific Computing '98* (1998), Bartz D., (Ed.), Springer, pp. 59–68.
11. Löffelmann H., Kucera T., Gröller E.: Visualizing poincaré maps together with the underlying flow. In *Mathematical Visualization. Proceedings of the International Workshop on Visualization and Mathematics '97* (1997), Hege H.-C., Polthier K., (Eds.), Springer, pp. 315–328.
12. MacKay R. S., Meiss J. D., Percival I. C.: Transport in hamiltonian systems. *Physica D 13D* (1984), 55–81.
13. Norbury J.: A family of steady vortex rings. *J. Fluid Mech. 57, Pt. 3* (1973), 417–431.
14. Saffman P. G.: *Vortex Dynamics*. Cambridge Univ. Press, Cambridge, UK, 1992.
15. Sil'nikov L. P.: A case of the existence of a denumerable set of periodic motions. *Sov. Math. Dokl. 6* (1965), 163–166.
16. Spohn A., Mory M., Hopfinger E.: Experiments on vortex breakdown in a confined flow generated by a rotating disc. *Journal of Fluid Mechanics 370* (1998), 73–99.
17. Sadlo F., Peikert R.: Topology-guided visualization of constrained vector fields. In *Proceedings of the 2005 Workshop on Topology-Based Methods in Visualization, Budmerice, Slovakia* (2007), p. (to appear).
18. Sotiropoulos F., Ventikos Y., Lackey T. C.. Chaotic advection in three dimensional stationary vortex-breakdown bubbles: Sil'nikov's chaos and the devil's staircase. *J. Fluid Mech. 444* (2001), 257–297.
19. Stalling D., Zöckler M., Hege H.-C.: Fast display of illuminated field lines. *IEEE Transactions on Visualization and Computer Graphics 3*, 2 (1997), 118–128.
20. Tricoche X., Garth C., Kindlmann G., Deines E., Scheuermann G., Ruetten M., Hansen C.: Visualization of intricate flow structures for vortex breakdown analysis. In *Proc. IEEE Visualization 2004* (October 2004), IEEE Computer Society, pp. 187–194.
21. Thompson M. C., Hourigan K.: The sensitivity of steady vortex breakdown bubbles in confined cylinder flows to rotating lid misalignment. *Journal of Fluid Mechanics 496* (Dec. 2003), 129–138.
22. Toth G.: The div b=0 constraint in shock-capturing magnetohydrodynamics codes. *Journal of Computational Physics 161* (2000), 605–652.

23. Theisel H., Weinkauf T., Hege H.-C., Seidel H.-P.: Saddle connectors - an approach to visualizing the topological skeleton of complex 3d vector fields. In *Proc. IEEE Visualization 2003* (Oct. 2003), pp. 225–232.
24. Weinkauf T., Hege H.-C., Noack B., Schlegel M., Dillmann A.: Coherent structures in a transitional flow around a backward-facing step. *Physics of Fluids 15*, 9 (September 2003), S3.
25. Wischgoll T., Scheuermann G.: Detection and visualization of closed streamlines in planar flows. *IEEE Transactions on Visualization and Computer Graphics 7*, 2 (2001), 165–172.
26. Ye X., Kao D., Pang A.: Strategy for scalable seeding of 3d streamlines. In *Proc. IEEE Visualization '05* (2005), pp. 471–478.

Bringing Topology-Based Flow Visualization to the Application Domain

Robert S. Laramee[1], Guoning Chen[2], Monika Jankun-Kelly[3], Eugene Zhang[2], and David Thompson[3]

[1] Department of Computer Science, Swansea University, Wales, UK
    r.s.laramee@swansea.ac.uk
[2] Oregon State University, Corvallis, Oregon
    {chengu,zhange}@eecs.oregonstate.edu
[3] Mississippi State University, Starkville, Mississippi
    {mjk,dst}@simcenter.msstate.edu

Summary. The visualization community is currently witnessing strong advances in topology-based flow visualization research. Numerous algorithms have been proposed since the introduction of this class of approaches in 1989. Yet despite the many advances in the field, topology-based flow visualization methods have, until now, failed to penetrate industry. Application domain experts are still, in general, not using topological analysis and visualization in daily practice. We present a range of state-of-the art topology-based flow visualization methods such as vortex core line extraction, singularity and separatrix extraction, and periodic orbit extraction techniques, and apply them to real-world data sets. Applications include the visualization of engine simulation data such as in-cylinder flow, cooling jacket flow, as well as flow around a spinning missile. The novel application of periodic orbit extraction to the boundary surface of a cooling jacket is presented. Based on our experiences, we then describe what we believe needs to be done in order to bring topological flow visualization methods to industry-level software applications. We believe this discussion will inspire useful directions for future work.

Key words: flow visualization, feature-based flow visualization, flow topology, applications

1 Introduction

Great progress has been made in the advancement of topology-based flow visualization methods since their introduction in 1989 [15]. Techniques for higher-order singularity extraction from vector fields have been introduced including a fast algorithm for 2D, steady-state data [43, 44]. Techniques for closed streamline extraction have been presented [52], including a grid-independent method [48]. The topology of unsteady flow can be extracted and visualized based on linear algebra [15], using streamline geometry [42], for 2D vector fields [50], and the tracking closed streamlines can be performed [53]. Several topology simplification algorithms have been written including multiresolution methods [8, 9], an area-based approach [10], topology-preservation-based compression of vector fields [35], and for vector field design [54]. Surface-based topology [16] such as detection of separation and attachment lines has been investigated [24, 26], and a fast version of the aforementioned topology [49]. And many vortex core line and vortex core region algorithms have been implemented [13], like the well-known λ_2 vortex core region-based method [19]. Levy et al. present an implementation that incorporates helicity [34]. Peikert and Roth describe and implementation that searches for local, parallel velocity and vorticity [39]. Roth and Peikert present an algorithm that extracts vortices of high-curvature [41]. The eigenvector method of Sujudi and Haimes [46], and the swirl parameter method of Berdahl and Thompson [4] both employ the notion that a vortex core line occurs in a region of complex eigenvalues where the velocity is parallel to the associated real eigenvector. Applications of vortex core line extraction to aerodynamics is described by Kenwright and Haimes [23, 25]. Vortex core line and region extraction techniques have also been developed for unsteady flow like the well-known predictor-corrector algorithm [1, 2] and an algorithm based on analysis of scalar values [3]. Reinders et al. present an application of tracking vortices trailing a tapered cylinder [40]. For a more complete overview of topology-based flow visualization research, we refer the readers to Laramee et al. [33].

Yet despite the many advances in topology-based analysis and visualization, this class of techniques is generally not used by domain experts in their daily routine. Functionality for the extraction and visualization of topology is not normally included in industry-level application software. The goal of this paper is to explore what needs to be done in order to bring topology-based methods to the application domain. We do so by drawing on our own experiences, namely, by applying topological methods to a collection of real-world data sets, identifying the insight they provide and at the same time, identifying the limitations of these approaches. We use topology-based extraction such as vortex core line extraction, singularity and separatrix extraction, and periodic orbit extraction techniques, in real engineering applications, including the visualization of engine simulation data such as in-cylinder flow, cooling jacket flow, as well as flow around a spinning missile. Our presentation includes the extraction of periodic orbits at the boundary surface of a cooling jacket– a novel application. We then describe what we believe needs to be done in order to bring topological flow visualization methods to industry-level software applications. We believe this discussion can play a role in steering future directions in the field.

We note that the current discussion fits in well with a larger trend in the visualization community. *Can visualization survive without customers?* [36] was the question posed by Bill Lorensen at the 2004 NIH/NSF workshop on visualization and research challenges [22]. Visualization research is not for the sake of visualization itself. In other words visualization is ultimately meant to help a user, i.e., someone normally outside the visualization community, gain insight into the problem they are trying to solve or the goal being sought after. Johnson called this problem "thinking about the science" [21]. Interdisciplinary collaboration can be very challenging. However, we do see signs of progress in this area. More quality, application-track papers have been published in recent years. We also note the emergence of the first Applied Visualization Conference (AppliedVis 2005) that took place in Asheville, North Carolina in April of 2005 (more information available at http://www.appliedvis.org). This topic was also the subject of recent panel discussions [14, 47] as well as a recent research paper [51].

2 Application: Simulation of In-Cylinder Flow

For flow entering and exiting a combustion chamber, the engineers responsible for the design try to create an ideal pattern of motion. The motion can be described as a swirling flow revolving around an imaginary, central axis residing inside the cylinder volume. One type of swirling motion, aptly called *swirl motion*, is depicted in Figure 3, left. The ideal swirl motion spirals around an axis aligned with the cylinder volume found at the center. Such an ideal is often strived for in diesel engines.

Another important pattern of flow is *tumble motion*, depicted in Figure 3, right. The axis of rotation in the tumble case is orthogonal to that of the swirl case. Also, the ideal motion is closer to a simple circle rather than a more spiral-like pattern. Since the axis of rotation is not aligned with the combustion chamber itself, this pattern of motion is more difficult to realize.

Achieving these ideal patterns of flow optimizes the mixture of oxygen and fuel during the ignition phase of the valve cycle. Optimal ignition leads to very desirable consequences associated with the combustion process including: more burnt fuel (less wasted fuel), lower emissions, and more output power.

2.1 Extraction and Visualization of Singularities and Separatrices

Extracting the singularities (or critical points) and related separatrices at the boundary of the geometry can provide valuable insight into the behavior of the flow directly, without the user having to search for the patterns of flow manually [15, 16, 17, 33]. Figure 4 shows the boundary topology, including singularities and separatrices for both cases of in-cylinder flow. In the case of the swirl motion associated with the diesel engine, the red separatrix at the boundary of the combustion chamber indicates a pattern of swirl motion consistent with the ideal shown in Figure 3, left [11, 32]. We can validate this to a certain extent by the addition of texture-based flow visualization, in this case using texture-advection approaches [30], to depict the characteristics of the entire flow at the boundary, not just the topological skeleton. In Figure 4 right, we see that the ideal pattern of tumble motion is not being realized. Instead of a single point-based recirculation zone directly in the middle of the combustion chamber, we see two dominant singularities: a saddle point in the upper, right-hand corner and a sink in the lower, left hand corner.

2.2 Future Direction: Extraction of Arbitrary Flow Patterns

Based on our experience of trying to extract the most important features from in-cylinder flow [6, 11, 31, 32], we can make the following observation: Tools capable of extracting more arbitrary patterns of flow motion, in 2D, 2.5D, or 3D would be very helpful to the engineering analysis community. (By 2.5D we mean surfaces in 3D.) For example, in the case of swirl motion, the ability to extract a 3D helical pattern directly would be very useful. Ideally, the user could specify an arbitrarily shaped curve and search the vector field according to this set of user-specified geometry (and topology). This idea is inspired by the fact that most geometries from automotive engineering have an ideal pattern of flow that the engineers are trying to realize when designing their models [11, 27, 28, 31, 32].

3 Application: Heat Transfer

The job of a cooling jacket is to transfer heat away from the engine block of an automobile [31]. The cooling jacket has an extremely complex geometry. The model grid consists of over 1.5 million unstructured, adaptive resolution tetrahedra, hexahedra, pyramid, and prism volume elements, the volume of which differs by more than six orders of magnitude. There are two main components to the ideal pattern of flow through a cooling jacket: a *longitudinal* motion lengthwise along the geometry and a *transversal* motion from cylinder block to head and from the intake to the exhaust side. These two components are sketched in Figure 1. The location of the inlet and outlet are also indicated. Any flow that deviates from this ideal, essentially the most efficient volume-filling path from inlet to outlet, results in less transfer of heat away from the engine block.

3.1 Periodic Orbit Detection on Boundary Surfaces

Figure 5, left shows a novel application–the extraction of periodic orbits (also known as closed streamlines) at the boundary surface of the cooling jacket using the algorithm of Chen et al. [7]. The algorithm automatically extracts and visualizes closed streamlines, 140 total in this example. In this application, periodic orbits are very relevant because they indicate areas of flow recirculation. Recirculation zones are very important to the engineers studying the design of this engine component because they detract from the goal of

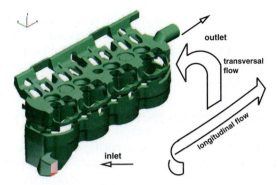

Fig. 1. The major components of the flow through a cooling jacket include a longitudinal component, lengthwise along the geometry and a transversal component in the upward-and-over direction. The inlet and outlet of the cooling jacket are also indicated

transferring heat away from the engine block. Hence, one of the goals of the engineer is to minimize the number of recirculation zones. Rather than having to manually inspect the surface for recirculation, the user can now extract this circular pattern of flow (tumble) directly.

Figure 5, right shows another very useful application of the periodic extraction algorithm, namely, to the boundary surface of a gas engine. In this visualization, we can see a large green periodic orbit hinting at a recirculation zone that corresponds very well to the ideal tumble motion depicted in Figure 3, right.

3.2 Future Direction: Higher Dimensional Topology Simplification

The complexity of the result in Figure 5, left clearly motivates the need for automatic or semi-automatic simplification algorithms for boundary topology. Hundreds of topological elements, both singularities and periodic orbits complicate the visualization result possibly adding undesirable noise. A filtering operation based on parameters such as (1) the size of periodic orbits, (2) the distance between neighboring singularities, (3) an error threshold or (4) some other size/distance metric would be helpful to the engineer in order to filter out some of the smaller scale (or larger) scale singularities. A similar statement can be made for the case of 3D topology [31].

3.3 Future Direction: Further Development of Topological Methods for Unsteady Flow

Practitioners typically deal with unsteady flow. Natural phenomena are time-dependent. Yet, topology-based flow visualization methods are still not fully understood in the context of time-dependent flow. For example, the interpretation of a separatrix for unsteady flow remains unclear. Separatrices are curves that segment the flow into different regions of asymptotic behavior. In the case of unsteady flow pathlines appear to be a natural choice (as opposed to streamlines) for separating the flow into regions of similar asymptotic behavior. In 3D, pathsurfaces appear to be a natural choice but very little work on pathsurfaces has been presented. Streaklines may also be a natural choice for separating unsteady flow in 2D, but what about 3D? Are "streaksurfaces" a better choice? To our knowledge, no work on streaksurfaces has been done. What kind of separating behavior is most relevant in the case of unsteady flow? What is the best approach to segmenting the different regions of flow in the case of unsteady flow? Questions such as these are only starting to be addressed [45].

Another problem lies in periodic orbit detection. Consensus lacks on whether periodic orbit detection makes more sense in the context of steady versus unsteady flow. One can argue that in fact, periodic orbit visualization is misleading in the case of steady-state flow based on the argument that no such paths exist in reality. This is because truly steady-state flow does not exist if we define steady-state flow as instantaneous in time. On the other hand, periodic orbits are unlikely to be detected in unsteady flow because spatio-temporal behavior would have to remain identical over all cycles. Singular orbit detection may make more sense in the context of unsteady flow. Further development in this direction would also make topology-based methods more appealing to practitioners.

4 Application: Spinning Missile

The spinning missile with dithering canards is representative of the type of complex geometry routinely used in vehicle performance simulations. Several properties of the spinning missile geometry contribute to the complexity of the vortical flow surrounding it [5]. The missile is at a three degree angle with respect to the supersonic incident flow. Fixed, canted tail fins cause the missile to spin about its longitudinal axis at a rate of 8.75hz (Figure 2, upper left). The missile's dithering canards provide pitch and yaw control by rotating about their attachment posts (Figure 2, lower right). To describe a single rotation of the missile, 360 time steps are employed. The canards complete several dither cycles during one missile rotation. The flow has some degree of periodicity, but due to the missile's nonzero angle of attack and spin, this period does not match the dither cycle. The simulation was performed on a mixed element mesh consisting of more than 35 million elements.

The engineers studying the missile were initially interested in whether the vortices coming off the canard tips impinged on the tail fins. Feature based vortex visualization was used to answer this question.

Fig. 2. Missile geometry: Canted tail fins (upper left) cause missile to spin about its longitudinal axis. Canards (lower right) rotate synchronously about axis passing through the missile body to provide pitch and yaw control

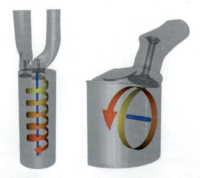

Fig. 3. (left) The *swirl* motion of flow in the combustion chamber of a diesel engine. Swirl is used to describe circulation about the axis aligned with the valve cylinder. The intake ports at the top provide the tangential component of the flow necessary for swirl. The data set consists of 776,000 unstructured, adaptive resolution grid cells. (right) Some in-cylinder flows require a *tumble* motion flow pattern in order to mix fuel with oxygen. Tumble flow circulates around an axis perpendicular to the cylinder axis, orthogonal to the case of swirl motion

4.1 Feature-Based Vortex Visualization

The vortex visualization method of Jankun-Kelly et al. [18] extracts vortex topology from computational fluid dynamics (CFD) data. Vortices can be topologically characterized by their core line and extent. The vortex core line is the curve about which streamlines swirl in a reference frame moving with the vortex. Please note this method uses **vortex topology**, not **vector field topology** (critical points, separatrices). This method exploits a technique to extract the vortex core lines from the line-type extrema of scalar fields and a

Fig. 4. (left) Swirl motion indicated by the helical separatrix at the boundary of the combustion chamber and (right) deviant tumble motion indicated by the boundary topology of the gas engine simulation For the singularities, green = source, red = sink, orange = attracting focus, cyan = repelling focus, blue = saddle. Red separatrices end at an attractor (sink). Green separatrices end at a repeller (source)

Fig. 5. (left) The flow topology, including 540 singularities and 140 periodic orbits extracted at the boundary surface of the cooling jacket. (right) Periodic orbits and singularities extracted at the boundary surface of the gas engine indicating tumble motion (green = source, red = sink, orange = attracting focus, cyan = repelling focus, blue = saddle)

novel k-means clustering algorithm to identify topologically complex vortical structures. The extent of a vortex is the boundary surface of the vortex core region [12]. Given core line and extent, additional vortex characteristics can be found. They include the sense of rotation, various measures of strength, and field values at the core.

This method has several strengths. Vortex detection is automatic rather than manual/interactive. Vortices can be extracted from practical engineering data that may be noisy, unstructured, and not resolved as well as we might like. Individual vortices of various strengths can be identified even in complex flows of multiple, interacting and merging vortices. Low level, large CFD data is distilled into compact, feature level vortex characteristic data such as core line, extent, and strength. These compact vortex detection results can be visualized interactively. All these properties of our method were requested by users.

4.2 Insight From Vortex Visualization

We have produced an animated visualization of the time varying vortical flow about the spinning missile. Several still images from the animation can be seen in Figure 6. Vortex core line and associated extent surface image pairs are shown from top to bottom, respectively. The color of each core line indicates rotation, clockwise or counterclockwise with respect to the local axial velocity. The extent surface is shaded by local tangential velocity, an indicator of vortex strength. This visualization allowed the engineers to answer their question: did canard tip vortices impinge on the tail fins and how strong

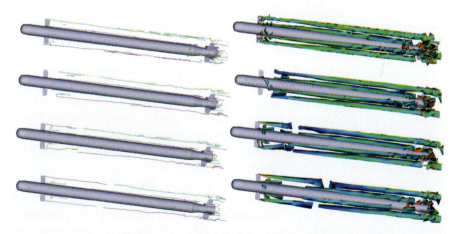

Fig. 6. Vortex core lines and extent for spinning missile with dithering canards for time steps t=726, 728, 730, and 732 (top to bottom). Purple indicates a left handed rotation while green indicates a right handed rotation. Alternate images show vortex extent with surface colored by the local tangential velocity. The scale on the non-dimensional tangential velocity is blue (0) to red (0.189)

were the vortices that did impinge. Other interesting vortex behavior was also observed. As the canards dithered, the strengths of their trailing vortices changed. This appears as a change in the shading of vortex extent. At certain roll angle/dither cycle combinations, each canard tip vortex changed its sense of rotation. This can be seen at time step 730 (Figure 6, third row from the top). The animation also revealed that vortices shed from the posts connecting the canards to the missile body traveled toward the tail fins. Engineers had originally been interested in canard tip vortices, but after seeing the vortex visualization animation, they decided that the post vortices merit further investigation. These insights were made possible by the topological feature extraction capabilities of this visualization method.

4.3 Future Direction: Noise Mitigation and Reduction

Noise is often present in the data we are given. We must work with the data we have, even when it is not quite ideal. This method is specifically designed to handle noisy data because doing so improves the quality of the visualization. More work needs to be done in the areas of noise reduction and mitigation The two most pressing problems caused by noisy data are noisy core lines and occasional C^0 core line discontinuity. For feature level applications, complete core lines are preferable to core line pieces. The vortex core line is needed to compute all other vortex characteristics. Less noise in the core line would make extent computation more robust. Therefore, improving core line quality would have a significant impact on the overall quality of the results.

4.4 Future Direction: Improved Extent Computation

Several challenges need to be overcome in order to make extent computation more accurate and robust. Extent outliers, which appear as jagged spikes or indentations, are visually jarring inaccuracies. They can be removed by smoothing the entire extent, but this adversely affects extent accuracy. More work is needed to identify and locally repair extent outliers. A better extent model is needed. The one currently used works well for isolated vortices but has more difficulty with multiple, interacting vortices.

Further, the extent computation needs to be made Galilean invariant, which could be accomplished by a shift of reference frame to one traveling with the vortex if the translational motion of the vortex were known. It should be noted that, for nonstationary vortices, the fluid velocity at the vortex core will not be the velocity of the vortex core, thereby making Galilean invariance difficult to achieve. These enhancements would facilitate application of the feature based vortex visualization method to a wider range of data.

4.5 Future Direction: Feature Level Verification by Determining Whether a Detected Feature is a Vortex

Feature level verification is needed to automatically eliminate false positives, a task currently done manually. Feature level verification would reduce visual clutter and make visualization more meaningful and easier to interpret. All methods that detect vortices by searching for line-type local extrema suffer from false positives. This is because local extrema are a necessary but not a sufficient condition for the existence of vortices. Swirling flow must also be present. After detecting possible vortex core lines, Jiang et al. verified vortices by checking for the presence of streamlines that made at least one complete revolution about the vortex core line [20]. The limitation of this vortex verification approach is that it is not Galilean invariant. Please note that the preceding vortex detection can be Galilean invariant.

4.6 Future Direction: Vortex Tracking for Unsteady Flow

This method could be further enhanced through the incorporation of vortex tracking. Vortex tracking would provide several benefits. Core line discontinuities, which should not be present, could be identified by comparing vortices at neighboring time steps. Vortex core velocity, needed for Galilean invariance, could be determined. Tracking could be another means of feature verification since vortices tend to persist over time, while some false positives are more transient. Tracking would permit rule mining, the discovery of rules governing vortical flow, leading to greater scientific insight [38].

5 Application Independent Directions

Here we describe some general future directions for bringing topology-based visualization to the industry domain, independent of a specific application.

5.1 Future Direction: More Accessible Implementations

In general, topological extraction methods are complex and difficult to implement. We believe that the success of some algorithms, e.g., Marching Cubes [37] or simple particle tracing is a result of an accessible implementation. Graceful and easier implementations for topological analysis, extraction, and visualization would help in bringing this class of methods to the domain experts. Alternatively, if the visualization community provided engineers with source code or pre-compiled software, they would be more likely to use our methods than in the case of having to implement from scratch. However, perhaps some methods can only be algorithmically simplified to a certain extent before they become ineffective. In general, users would rather not re-implement methods. They prefer off the shelf software with an easy to use and intuitive UI.

5.2 Future Direction: Faster Computation Time

Clearly one aspect that would make topology-based flow visualization more attractive to practitioners is faster computation. In general, literature on feature-based flow visualization does not report performance times [33]. If performance times are not reported, there is usually a reason. Topology-based methods that are fast, or even interactive, would certainly enhance their attractiveness to those outside the visualization community.

6 Summary and Conclusions

Despite the recent advances in topology-based flow visualization research, topological methods are still generally not found in commercial software packages. We have presented a range of topological analysis and visualization tools and their application to real-world problems. Included is a novel application–the extraction and visualization of periodic orbits at the boundary surface of a cooling jacket. Drawing on this experience, we presented a list of future work in this area necessary in order to bring topology-based flow visualization to the application domain. We have identified the following tasks:

1. Extraction of Arbitrary Flow Patterns: the ability to extract and visualize user-defined patterns from flow
2. Higher Dimensional Topology Simplification: methods that can either automatically or semi-automatically simplify topology on boundary surfaces and in 3D
3. More Accessible Implementations: implementations of extraction methods which are easier and more accessible to the engineers that must write them
4. Faster Performance: implementations that are interactive, or nearly interactive, would be ideal
5. Further Development of Topological Methods for Unsteady Flow: a greater understanding of topology-based methods in the context of unsteady flow is needed
6. Noise Reduction and Mitigation: the development of topology-based methods which are less sensitive to noisy data
7. Improved Extent Computation: accurate methods to handle outliers in the data
8. Feature Level Verification: automatic methods for the elimination of false positives
9. Vortex Core Tracking for Unsteady Flow: including a formal definition of a vortex
10. Improved Dissemination: a better transfer of knowledge from the visualization community is necessary.

Lack of communication between communities is also a problem. Currently there is quite a gap between the visualization research community and prospective users. In fact, other communities such as the engineering analysis community [29]. are not even aware that a visualization community exists. Visualization scientists need to explain what tools and techniques are available and how they can be used to solve problems in science and engineering. Practitioners could also explain why they would like to visualize their data and what questions they're trying to answer. Closely related is the lack of inter-community knowledge transfer and a lack of educational literature.

We believe this discussion may play a role in steering future work in this field to the point where topological methods may even be included in industry-grade software. Certainly, a lot of work remains for topology-based methods to spread beyond the visualization community.

Acknowledgements

We would like to thank Ming Jiang, Raghu Machiraju, Christoph Garth, and Helwig Hauser for their valuable contributions. We also wish to thank Konstantin Mischaikow and Greg Turk for their valuable suggestions. We especially thank Harry Yeh in the Coastal and Ocean Engineering Department at Oregon State University for his very valuable user feedback on the paper.

References

1. D. C. Banks and B. A. Singer. Vortex Tubes in Turbulent Flows: Identification, Representation, Reconstruction. In *Proceedings IEEE Visualization '94*, pages 132–139, October 1994.
2. D. C. Banks and B. A. Singer. A Predictor-Corrector Technique for Visualizing Unsteady Flow. *IEEE Transactions on Visualization and Computer Graphics*, 1(2):151–163, June 1995.
3. D. Bauer and R. Peikert. Vortex Tracking in Scale-Space. In *Proceedings of the Symposium on Data Visualisation 2002*, pages 233–240. Eurographics Association, 2002.
4. C. H. Berdahl and D. S. Thompson. Eduction of Swirling Structure Using the Velocity Gradient Tensor. *American Institute of Aeronautics and Astronautics (AIAA) Journal*, 31(1):97–103, January 1993.
5. E. L. Blades and D. L. Marcum. Numerical Simulation of a Spinning Missile with Dithering Canards Using Unstructured Grids. *Journal of Spacecraft and Rockets*, 41(2):248–256, 2004.
6. G. Chen, R. S. Laramee, and E. Zhang. Advanced Visualization of Engine Simulation Data Using Texture Synthesis and Topological Analysis. In *NAFEMS World Congress Conference Proceedings*. NAFEMS–The International Association for the Engineering Analysis Community, May 22–25 2007. (full proceedings on CDROM).
7. G. Chen, K. Mischaikow, R. S. Laramee, P. Pilarczyk, and E. Zhang. Vector Field Editing and Periodic Orbit Extraction Using Morse Decomposition. *IEEE Transactions on Visualization and Computer Graphics*, 13(4):769–785, Jul/Aug 2007.

8. W. de Leeuw and R. van Liere. Visualization of Global Flow Structures Using Multiple Levels of Topology. In *Data Visualization '99 (VisSym '99)*, pages 45–52. May 1999.
9. W. de Leeuw and R. van Liere. Multi-level Topology for Flow Visualization. *Computers and Graphics*, 24(3):325–331, June 2000.
10. W. C. de Leeuw and R. van Liere. Collapsing Flow Topology Using Area Metrics. In *Proceedings IEEE Visualization '99*, pages 349–354, 1999.
11. C. Garth, R.S. Laramee, X. Tricoche, J. Schneider, and H. Hagen. Extraction and Visualization of Swirl and Tumble Motion from Engine Simulation Data. In *Topology-Based Methods in Visualization (Proceedings of Topo-in-Vis 2005)*, Mathematics and Visualization, pages 121–135. Springer, 2007.
12. C. Garth, X. Tricoche, T. Salzbrunn, T. Bobach, and G. Scheuermann. Surface Techniques for Vortex Visualization. In *Data Visualization, Proceedings of the 6th Joint IEEE TCVG–EUROGRAPHICS Symposium on Visualization (VisSym 2004)*, pages 155–164, May 2004.
13. R. Haimes and D. Kenwright. On the Velocity Gradient Tensor and Fluid Feature Extraction. Technical Report AIAA Paper 99–3288, American Institute of Aeronautics and Astronautics, 1999.
14. H. Hauser, P.T. Bremer, H Theisel, M. Trener, and X. Tricoche. Panel: What are the most demanding and critical problems, and what are the most promising research directions in Topology-Based Flow Visualization? In *Topology-Based Methods in Visualization Workshop, 2005*, September 2005. Held in Budmerice, Slovakia.
15. J. L. Helman and L. Hesselink. Representation and Display of Vector Field Topology in Fluid Flow Data Sets. *IEEE Computer*, 22(8):27–36, August 1989.
16. J. L. Helman and L. Hesselink. Surface Representations of Two- and Three-Dimensional Fluid Flow Topology. In *Proceedings IEEE Visualization '90*, pages 6–13, 1990.
17. J. L. Helman and L. Hesselink. Visualizing Vector Field Topology in Fluid Flows. *IEEE Computer Graphics and Applications*, 11(3):36–46, May 1991.
18. M. Jankun-Kelly, M. Jiang, D. Thompson, and R. Machiraju. Vortex Visualization for Practical Engineering Applications. *IEEE Transactions on Visualization and Computer Graphics*, 12(5):957–964, 2006.
19. J. Jeong and F. Hussain. On the Identification of a Vortex. *Journal of Fluid Mechanics*, 285:69–94, 1995.
20. M. Jiang, R. Machiraju, and D. S. Thompson. Geometric Verification of Swirling Features in Flow Fields. In *Proceedings of IEEE Visualization '02*, pages 307–314, 2002.
21. C. R. Johnson. Top Scientific Visualization Research Problems. *IEEE Computer Graphics and Applications*, 24(4):13–17, July/August 2004.
22. C. R. Johnson, R. Moorehead, T. Munzner, H Pfister, P. Rheingans, and T. S. Yoo. NIH/NSF Visualization Research Challenges (Final Draft, January 2006). Technical report, 2006.
23. D. Kenwright and R. Haimes. Vortex Identification–Applications in Aerodynamics. In *Proceedings IEEE Visualization '97*, pages 413–416, November 1997.
24. D. N. Kenwright. Automatic Detection of Open and Closed Separation and Attachment Lines. In *Proceedings IEEE Visualization '98*, pages 151–158, 1998.
25. D. N. Kenwright and R. Haimes. Automatic Vortex Core Detection. *IEEE Computer Graphics and Applications*, 18(4):70–74, July/August 1998.

26. D. N. Kenwright, C. Henze, and C. Levit. Features Extraction of Separation and Attachment Lines. *IEEE Transactions on Visualization and Computer Graphics*, 5(2):135–144, 1999.
27. R. S. Laramee. Effective Visualization of Heat Transfer. In *The 12th International Symposium on Flow Visualization (ISFV12)*, September 10–24 2006. (proceedings on CDROM).
28. R. S. Laramee and C. Garth. Stream Surfaces and Texture Advection: A Hybrid Metaphor for Visualization of CFD Simulation Results. In *The 12th International Symposium on Flow Visualization (ISFV12)*, September 10–24 2006. (proceedings on CDROM).
29. R. S. Laramee and H. Hauser. Interactive 3D Flow Visualization Using Textures and Geometric Primitives. In *NAFEMS World Congress Conference Proceedings*, page 75. NAFEMS–The International Association for the Engineering Analysis Community, May 17–20 2005. (full proceedings on CDROM).
30. R. S. Laramee, J. J. van Wijk, B. Jobard, and H. Hauser. ISA and IBFVS: Image Space Based Visualization of Flow on Surfaces. *IEEE Transactions on Visualization and Computer Graphics*, 10(6):637–648, November 2004.
31. R. S. Laramee and C. Ware. Visual Interference with a Transparent Head Mounted Display. In *CHI 2001, Conference on Human Factors in Computing Systems, Extended Abstracts*, pages 323–324. ACM SIGCHI, ACM Press, 31 March–5 April 2001.
32. R. S. Laramee, D. Weiskopf, J. Schneider, and H. Hauser. Investigating Swirl and Tumble Flow with a Comparison of Visualization Techniques. In *Proceedings IEEE Visualization 2004*, pages 51–58, 2004.
33. R. S. Laramee, H. Hauser, L. Zhao, and F. H. Post. Topology Based Flow Visualization: The State of the Art. In *Topology-Based Methods in Visualization (Proceedings of Topo-in-Vis 2005)*, Mathematics and Visualization, pages 1–19. Springer, 2007.
34. Y. Levy, D. Degani, and A. Seginer. Graphical Visualization of Vortical Flows by Means of Helicity. *AIAA Journal*, 28:1347–1352, 1990.
35. S. K. Lodha, J. C. Renteria, and K. M. Roskin. Topology Preserving Compression of 2D Vector Fields. In *Proceedings IEEE Visualization 2000*, pages 343–350, 2000.
36. B. Lorensen. Panel Statement: On the Death of Visualization: Can It Survive Without Customers? In *NIH/NSF Fall 2004 Workshop on Visualization Research Challenges*, September 2004. Available for download: http://visual.nlm.nih.gov/.
37. W. E. Lorensen and H. E. Cline. Marching Cubes: a High Resolution 3D Surface Construction Algorithm. In *Computer Graphics (Proceedings of ACM SIGGRAPH 87, Anaheim, CA)*, pages 163–170. ACM, July 27–31 1987.
38. R. Machiraju, S. Parthasarathy, J. W. Wilkins, D. S. Thompson, B. Gatlin, D. A. Richie, T.-S. Choy, M. Jiang, S. Mehta, M. Coatney, S. A. Barr, and K. Hazzard. Mining Temporally-Varying Phenomena in Scientific Datasets. In *Data Mining: Next Generation Challenges and Future Directions*, pages 273–290. AAAI/MIT Press, 2004.
39. R. Peikert and M. Roth. The Parallel Vectors Operator - A Vector Field Visualization Primitive. In *Proceedings of IEEE Visualization '99*, pages 263–270. IEEE Computer Society, 1999.

40. F. Reinders, I. A. Sadarjoen, B. Vrolijk, and F. H. Post. Vortex Tracking and Visualisation in a Flow Past a Tapered Cylinder. In *Computer Graphics Forum*, volume 21(4), pages 675–682. November 2002.
41. M. Roth and R. Peikert. A Higher-Order Method For Finding Vortex Core Lines. In *Proceedings IEEE Visualization '98*, pages 143–150, 1998.
42. I. A. Sadarjoen and F. H. Post. Detection, Quantification, and Tracking of Vortices using Streamline Geometry. *Computers and Graphics*, 24(3):333–341, June 2000.
43. G. Scheuermann, H. Hagen, H. Krüger, M. Menzel, and A. Rockwood. Visualization of Higher Order Singularities in Vector Fields. In *Proceedings IEEE Visualization '97*, pages 67–74, October 1997.
44. G. Scheuermann, H. Krüger, M. Menzel, and A. P. Rockwood. Visualizing Nonlinear Vector Field Topology. *IEEE Transactions on Visualization and Computer Graphics*, 4(2):109–116, April/June 1998.
45. K. Shi, H. Theisel, T. Weinkauf, H. Hauser, H.-C. Hege, and H.-P. Seidel. Path Line Oriented Topology for Periodic 2D Time-Dependent Vector Fields. In *Data Visualization, The Joint Eurographics-IEEE VGTC Symposium on Visualization (EuroVis 2006)*, pages 139–146, 2006.
46. D. Sujudi and R. Haimes. Identification of Swirling Flow in 3D Vector Fields. Technical Report AIAA Paper 95–1715, American Institute of Aeronautics and Astronautics, 1995.
47. H. Theisel, T. Ertl, H.-C. Hagen, B. Noack, and G. Scheuermann. Panel: Why are topological methods not included in commercial visualization systems? In *Topology-Based Methods in Visualization Workshop, 2005*, September 2005. Held in Budmerice, Slovakia.
48. H. Theisel, T. Weinkauf, H.-P. Seidel, and H. Seidel. Grid-Independent Detection of Closed Stream Lines in 2D Vector Fields. In *Proceedings of the Conference on Vision, Modeling and Visualization 2004 (VMV 04)*, pages 421–428, November 2004.
49. X. Tricoche, C. Garth, and G. Scheuermann. Fast and Robust Extraction of Separation Line Features. In *Proceedings of Seminar on Scientific Visualization 2003, Schloss Dagstuhl*, 2003.
50. X. Tricoche, G. Scheuermann, and H. Hagen. Topology-Based Visualization of Time-Dependent 2D Vector Fields. In *Proceedings of the Joint Eurographics - IEEE TCVG Symposium on Visualization (VisSym '01)*, pages 117–126, May 28–30 2001.
51. J. J. van Wijk. The Value of Visualization. In *Proceedings IEEE Visualization '05*, pages 79–86. IEEE Computer Society, 2005.
52. T. Wischgoll and G. Scheuermann. Detection and Visualization of Closed Streamlines in Planar Fields. *IEEE Transactions on Visualization and Computer Graphics*, 7(2):165–172, 2001.
53. T. Wischgoll, G. Scheuermann, and H. Hagen. Tracking Closed Streamlines in Time Dependent Planar Flows. In *Proceedings of the Vision Modeling and Visualization Conference 2001 (VMV 01)*, pages 447–454, November 21–23 2001.
54. E. Zhang, K. Mischaikow, and G. Turk. Vector field design on surfaces. *ACM Transactions on Graphics*, 25(4):1294–1326, 2006.

Computing Center-Lines: An Application of Vector Field Topology

Thomas Wischgoll

Wright State University, `thomas.wischgoll@wright.edu`

Summary. Curve-skeletons of 3-D objects are medial axes shrunk to a single line. There are several applications for curve-skeletons. For example, animation of 3-D objects, such as an animal or a human, as well as planning of flight paths for virtual colonoscopy. Other applications are the extraction of center lines within blood vessels where center lines are used to quantitatively measure vessel length, vessel diameter, and angles between vessels. The described method computes curve-skeletons based on a vector field that is orthogonal to the object's boundary surface. A topological analysis of this field then yields the center lines of the curve-skeletons. In contrast to previous methods, the vector field does not need to be computed for every sampled point of the entire volume. Instead, the vector field is determined only on the sample points on the boundary surface of the objects. Since most of the computational time was spent on calculating the force field in previous methods, the proposed approach requires significantly less time compared to previous vector-based techniques while still achieving a better accuracy and robustness compared to methods based on Voronoi tessellations.

1 Introduction

Curve-skeletons describe the very basic features of an object. They describe a thinned version of the object represented as some type of stick model resulting in the center-lines of the object. Therefore, the use of curve-skeletons can prove useful for applications, such as animation [43] or flight planning for virtual colonoscopy [19]. Similarly, accurate curve-skeleton methods can be used for extracting quantitative measurements from computed tomography (CT) scanned vascular structures. Here, the curve-skeleton describes the center lines of the vessels. These can then be used to measure vessel radius, vessel lengths, and angles between vessels within the volumetric data set retrieved by using the CT scanner. This is the application that motivated the development of the algorithm described in this paper. In order to derive these measurements from the volumetric data set, an accurate extraction method for curve-skeletons is desirable. For example, thinning-based techniques that work in the voxel space of the volumetric data set tend to generate jagged lines which are in no way suitable for determining angles between vessels. Similarly, inaccuracies can occur when computing the radii of the vessels. Hence, an approach that only uses the volumetric data set in order to identify the boundary surface of the contained object is more promising.

The algorithm described in this paper is exactly of this type. It is capable of extracting the boundary surface of an object that is defined by a volumetric data set at sub-voxel level. For this, it determines the location of the maximal gradient within the volumetric data set similar to Canny's [9] maxima-suppression technique but extended to three dimensions. Since the algorithm only relies on points extracted from the volumetric data set but not on its underlying structured grid, it can also be applied to objects defined by a point set without any restrictions.

Techniques used for computing the topological graph of a vector field are applied to determine the curve-skeleton. First, for all points on the object's boundary vectors are computed that are orthogonal to the boundary surface. There are different options for computing these vectors. They either can be derived by determining the normal vector of a plane that is defined by a least-square fit of the point and its neighbors. Or – in case of the object being defined by a volumetric data set – the image gradients determined in the previous step can be used. In both cases, the normal vectors can be determined in such a way that they are facing inwards with respect to the object. The entire vector field can then be determined by computing a tetrahedrization of the entire point cloud and then linearly interpolating within the tetrahedra. In order to ensure that only the curve-skeleton inside the object is extracted, all tetrahedra that are located outside the object are removed based on the normal vectors.

A topological analysis of the vector field within the faces of every tetrahedron yields points on the curve-skeleton. By following the topology of the tetrahedrization, points on the curve-skeleton within neighboring tetrahedra can be connected resulting in the entire curve-skeleton.

A detailed description of the algorithm can be found in section 3. The next section illustrates related work and compares it to the described approach. Subsequently, the theoretical background with regard to the topological analysis of vector fields is explained. Section 4 shows results of the algorithm applied to various data sets, followed by conclusions and future work.

2 Related Work

Several approaches for extracting curve-skeletons or medial axes can be found in the literature. A very good overview of available techniques can be found in the paper by Cornea et al. [11].

Some methods start with all voxels of a volumetric data set and use a thinning technique to shrink down the object to a single line. Directional thinning approaches use a specific order in which voxels are removed. For example, directions, such as up or down, are used to define this order and conditions are used to identify endpoints [3, 8, 18, 23, 25, 30, 31, 34, 42]. Since these methods are sensitive with respect to the order in which the voxels are removed the resulting curve-skeleton may not be centered. Non-directional methods [6, 39] or fully parallel approaches [14, 27, 29] do not suffer from this disadvantage. Ideally, the topology of the object should be observed. Such an approach was proposed by Lobregt et al. [24] which is the basic technique used in commercial software systems, such as Analyze^{TM} developed by the Mayo Clinic. The disadvantage of this approach is that it tends to produce jagged lines which do not allow accurate measurements of angles between parts of the object, such as individual vessels of a vascular structure. Other approaches [40] classify the voxels in different groups, such as edge, inner, curve, or junction and re-classify after removal of a voxel. A similar algorithm is proposed by Palagyi et al. [30]. The disadvantage of thinning algorithms is that they can only be applied to volumetric data sets due to the nature of these algorithms.

To avoid this disadvantage, other approaches deploy the distance transform [17] or distance field in order to obtain a curve-skeleton. For each point inside the object, the smallest distance to the boundary surface is determined. For this, the Euclidian metric or the $<3,4,5>$ metric [5] can be used. Also, fast marching methods [36, 41] can be deployed to compute the distance field. Voxels representing the center lines of the object are identified by finding ridges in the distance field. The resulting candidates must then be pruned first. The resulting values are then connected using a path connection or minimum span tree algorithm [38, 44, 48]. Methods used to identify points on the ridges include distance thinning [10, 15, 16, 33], divergence computing [7], gradient searching [4], thresholding the bisector angle [28], geodesic front propagation [32], or shrinking the surface along the gradient of the distance field [35]. The distance field can also be combined with a distance-from-source field to compute a skeleton [49]. Based on an anisotropic diffusion applied to the image gradients, Yu et al [47] extract skeletons from 2D images.

Techniques based on Voronoi diagrams [2, 13] define a medial axis using the Voronoi points. Since this approach usually does not result in a single line but rather a surface shaped object, the points need to be clustered and connected in order to obtain a curve-skeleton. Voronoi-based methods can be applied to volumetric data sets as well as point sets. Due to the fact that clustering of the resulting points is required these approaches can lack some accuracy.

3 Methodology

It is assumed that the reader is familiar with singularities in vector fields and 2-D vector field topology. If necessary, a good overview of these topics can be found in [20, 45, 46].

The algorithm for determining the curve-skeleton consists of several steps. If the object is given as a volumetric data set the object's boundary has to be extracted first. Then, a vector field is computed that is orthogonal to the object's boundary surface. Once the vector field is computed, the curve-skeleton can be determined by applying a topological analysis to this vector field. The following subsections explain these steps in detail.

3.1 Extracting the Boundary of the Object

If the object is given as a set of points, for example measured by a laser scanner, the object's boundary is already defined. If the object is defined by a volumetric data set, for example from a CT scan, the boundary of the object has to be determined first. A volumetric data set consists of voxels aligned along a regular, three-dimensional grid. Since it is generally not likely that the boundary of the original object is exactly located at these voxels, better precision can be achieved by finding the exact location between a set of voxels. Since an accurate representation of the object's boundary is crucial to the algorithm, improving the precision is an essential step. The method used within the described system uses similar techniques as described by Canny's non-maxima suppression [9, 21] but extended to three dimensions.

First, the image gradients are computed. Using a fixed threshold, all voxels with a gradient length below this threshold are neglected. Then, the gradients of the remaining voxels are compared to its neighbors to identify local maxima along the gradient. In 3-D, the direct neighborhood of a single voxel generally consists of 26 voxels forming a cube that surrounds the current voxel. In order to find the local maximum along the current gradient, the gradients in the neighborhood in positive and negative direction of the current gradient have to be determined. The current implementation of the described system uses tri-linear interpolation. Based on these three gradient values, i.e. the original image gradient and the two interpolated values, the location of the maximal gradient can be identified, resulting in a more accurate representation of the object's boundary and therefore a more precise center line. Hence, this step improves the accuracy of the resulting curve-skeleton. However, the algorithm would still work using the original voxels identified.

Once all points on the boundary are extracted from the volumetric data set using this gradient approach with sub-voxel precision, the resulting point cloud can be further processed in order to identify the skeleton.

3.2 Computing the Vector Field

The described method computes a curve-skeleton by applying a topological analysis to a vector field that is determined based on the geometric configuration of the object of which the curve-skeleton is to be determined. The vector field is computed in such a way that the vectors are orthogonal to the object's boundary surface. The vectors inside the object are then interpolated linearly.

Different approaches are possible for calculating such a vector field. A repulsive force field can be determined that uses the surrounding points on the object's boundary surface as used by Cornea et al. [12]. The repulsive force is defined similarly to the repulsive force of a generalized potential field [1, 22]. The basic idea is to simulate a potential field that is generated by the force field inside the object by charging the object's boundary. Another way is to define a normal vector by using the neighboring points in addition to the current one and then defining a plane that is approximated by these points. The normal of this plane then defines the vector corresponding to the current point.

If the data is given as a volumetric data set the image gradients can be used to define the vectors on the object's boundary surface. These image gradients are computed already in the previous step as they are needed for extracting the object's boundary and determining the sub-voxel precision. The image gradients are computed using the derivative of a Gaussian which also results in a smoothing of the gradient vectors to address noise that may be present in the data. In our experiments, the resulting gradient vectors were sufficiently accurate to determine an accurate center line as shown by the validation of the algorithm. Note that all three methods result in vectors pointing to the inside of the object.

3.3 Determining the Curve-Skeleton

In order to determine the curve-skeleton of the object, a tetrahedrization of all points on the object's boundary is computed first. For this, Si's [37] very fast implementation of a Delaunay tetrahedrization algorithm is used. By using the previously computed vectors that point to the inside of the object, outside tetrahedra can be distinguished from tetrahedra that are located inside the object. This way, all outside tetrahedra can be removed, leaving a Delaunay tetrahedrization of the inside of the object only. Since vectors are known for each vertex of every tetrahedron, the complete vector field can be computed using this tetrahedrization by interpolating linearly within each tetrahedron. This vector field is then used to identify points of the curve-skeleton which are connected which each other later on.

Since the vector field is now defined within the entire object, one could use an approach similar to the one used by Cornea et al. [12] at this point and compute the 3-D topological skeleton of the vector field which yields the curve-skeleton of the object. However, since singularities are very rare in a 3-D vector field Cornea et al. had to introduce additional starting points for the separatrices, such as low divergence points and high curvature points, in order to get a good representation of the curve-skeleton. Therefore, a different approach is described in this paper that analyzes the vector field on the faces of the tetrahedra. In order to be able to perform a topological analysis on the faces of the tetrahedra, the vector field has to be projected onto those faces first. Since linear interpolation is used within the tetrahedra, it is sufficient to just project the vectors at the vertices onto each face and then interpolate linearly within the face using these newly computed vectors. Based on the resulting vector field, a topological analysis can be performed on each face of every tetrahedron.

Points on the curve-skeleton can then be identified by computing the singularities within the vector field interpolated within each and every face of the tetrahedrization. For example, for a perfectly cylindrical object, the vectors computed at the cylinder's boundary point directly at the center of the cylinder. When looking at the resulting vector field at a cross section of the cylinder, a focus singularity is located at the center of the cylinder within this cross section. The location of this focus singularity resembles a point on the curve-skeleton of the cylinder. Hence, a singularity within a face of a tetrahedron resembles a point of the curve-skeleton. Since the vectors at the object's boundary point inwards, only sinks need to be considered in order to identify the curve-skeleton. Due to the fact that not all objects are cylindrical in shape and due to numerical errors and tolerances, points on the curve-skeleton can be identified by looking for sinks that resemble focus and spiral singularities.

Obviously, only faces that are close to being a cross section of the object should be considered to identify points on the curve-skeleton. In order to determine tetrahedra whose faces resemble a cross-section of the object, the vectors at the vertices can be used. If the vectors at the vertices, which are orthogonal to the object's boundary, are approximately coplanar with the face, then this face describes a cross section of the object. As a test, the scalar product between the normal vector of the face and the vector at all three vertices can be used. If the result is smaller than a user-defined threshold this face is used to determine points on the curve-skeleton. A fixed threshold works for most data sets. However, in some cases a sub-optimal choice of this threshold can result in false bifurcations where small segments (usually just a single line segment) appears to be branching off of the main skeleton. Computing the singularity on one of the faces fulfilling the threshold criterion then results in a point which is part of the curve-skeleton. Since linear interpolation is used within the face, only a single singularity can be present in each face.

Once individual points of the curve-skeleton are computed by identifying the focus and spiral singularities within the faces of the tetrahedra, this set of points needs to be connected in order to retrieve the entire curve-skeleton. Since the tetrahedrization describes the topology of the object, the connectivity information of the tetrahedra can be used. Thus, identified points of the curve-skeleton of neighboring tetrahedra are connected with each other forming the entire curve-skeleton. In some occurrences, an additional gap closing step is required. Particularly, this is sometimes necessary at bifurcations where a smaller vessel branches off of a comparably large one. An example can be seen in figure 1 where the connection between the major vessel and the vessel branching off at the center of the image is missing. This can be easily detected by looking at the terminal nodes of the graph representing the center lines. For all these nodes, the algorithm checks if there is another graph it can

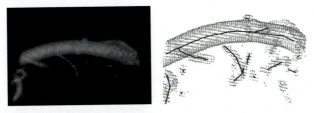

Fig. 1. Sub-section of the porcine heart data set visualized as a volume rendered image (left) and the extracted curve-skeleton of the same sub-section of the porcine heart data set (right)

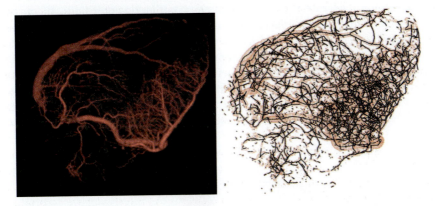

Fig. 2. Volume rendering (with shading enabled) of a previously perfused porcine heart which was scanned using a standard hospital CT scanner (left) and curve-skeleton of the porcine heart data set using the described algorithm before the gap closing step; to enhance visibility in the paper the thickness of the lines and points within the volumetric data set is increased (right)

be connected to. If this connection is within a vessel, the connection is added. To check this, the tetrahedrization can be used since the connection is only inside a vessel if it is entirely covered by tetrahedra.

4 Results

The algorithm was tested on several different data sets. It was mainly designed for extracting center lines from CT scanned volumetric data sets of porcine hearts where the arterials were previously perfused with a contrast enhancing polymer and computing the vessel radii as the distance between the center line and the vessel boundary. Figure 2(left) shows an example of such a data set.

The described algorithm is capable of extracting the curve-skeleton from such a volumetric data set in order to identify the center lines of the arterial vessels. The resulting curve-skeleton is depicted in figure 2(right). The figure shows the curve-skeleton extracted by the described algorithm before the gap closing step was applied as well as the point set representing the vessel boundaries.

Due to the densely located vessels of the right coronary arterial (RCA) tree, the extracted curve-skeleton seems rather cluttered and it is difficult to distinguish between lines at different depths. However, the extracted curve-skeleton describes the center lines of the arterial vessels found within the data set very well as illustrated below. When using a sub-section of the porcine heart data set, it can be seen that the curve-skeleton is located at the center of the arterial vessels, as shown in figure 3.

In order to validate the accuracy of the computed center lines, the vessel radii were computed for the main trunk of the arterial branches of a series of five porcine heart data sets as the distance between the center line and the vessel boundary and then compared to manual optical measurements. The optical measurements were performed by a domain expert after digesting away the tissue leaving only the contrast enhancing polymer used to fill the arteries. Based on the cast formed by the polymer, radii and distance to the most proximal vessel were then measured. Figure 4 shows a comparison of

Fig. 3. Sub-section of the porcine heart data set showing the extracted center line (left) and the underlying tetrahedrization used for identifying the center line (right)

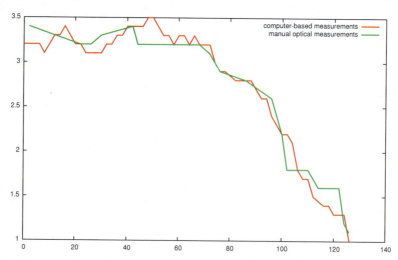

Fig. 4. Comparison of computed vessel radii (red) and manual optical measurements (green) for a typical specimen

the two different measurements for a typical data set. The agreement between the measurements is very good, with an error of 0.06mm (scan resolution was 0.6mm), which underlines the accuracy of the center lines. For the three major branches (LAD, LCX, and RCA) of the five procine hearts, the root mean square error between the two measurements is 0.16mm and the average deviation is 0.13mm.

The described algorithm for extracting curve-skeletons has some definite advantages over, for example, Voronoi-based approaches. Voronoi-based approaches define a medial axis that is not necessarily a line, but more like a set of points defining a surface. In a post-processing step, these points need to be shrunk down to define a line. Even though the arterial vessels are rather round due to the fact that they are pressurized, a Voronoi-based algorithm only determines a fuzzy line around the actual center line of the vessels. In addition, Voronoi-based approaches tend to generate significantly more false bifurcations in form of small, short vessels branching off. The root mean square error of the measurements computed using the presented technique of 0.16mm are more precise compared to other techniques found in the literature[26], where the root mean square error ranges from 0.2mm to 0.6mm with similar scan resolutions. The performance of the algorithm was compared to another vector-based algorithm by Conrea et al.[11]. The computation of the potential field as the first step of that algorithm would have taken several months. The presented algorithm requires only little more than an hour for the entire analysis of the data set.

The described algorithm works well with other types of data sets. The first example is a pure cylindrical shaped data set. The cylinder is perfectly round; hence, the algorithm should find a straight line as the curve-skeleton. As can be seen in figure 5(a), the algorithm generates the correct curve-skeleton for this data set exactly. The tetrahedrization of a small slice shown on the right

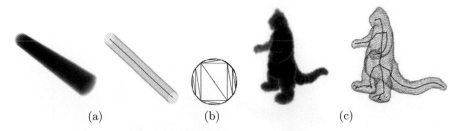

Fig. 5. Cylinder data set visualized as a volume rendered image (a) and the extracted curve-skeleton of the same data set (center); to enhance visibility in the paper the thickness of the lines and points within the image is increased. To illustrate the algorithm, the tetrahedrization within a small slice is shown as well (b). Monster data set (c) visualized as a volume rendered image (left) and the extracted curve-skeleton of the same data set (right); to enhance visibility in the paper the thickness of the lines and points within the image is increased

illustrates the principles of the algorithm. There are two large triangles cutting through the cylinder with which a point on the center line is identified marked with a dot at the center of the image.

Figure 5(c) depicts the results of the next example, the monster data set. Again, both the volume rendered image and the curve-skeleton including the point describing the object's boundary are shown. This example shows that the algorithm works well even with non-tubular objects. More examples can be found in figures 6 to 9.

Fig. 6. Cylinder data set visualized as a volume rendered image (left) and the extracted curve-skeleton of the same data set (right); to enhance visibility in the paper the thickness of the lines and points within the image is increased

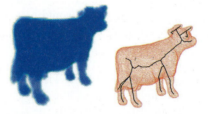

Fig. 7. Cow data set visualized as a volume rendered image (left) and the extracted curve-skeleton of the same data set (right); to enhance visibility in the paper the thickness of the lines and points within the image is increased

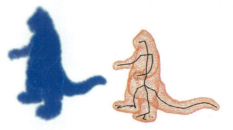

Fig. 8. Monster data set visualized as a volume rendered image (left) and the extracted curve-skeleton of the same data set (right); to enhance visibility in the paper the thickness of the lines and points within the image is increased

Fig. 9. Mushroom data set visualized as a volume rendered image (left) and the extracted curve-skeleton of the same data set (right); to enhance visibility in the paper the thickness of the lines and points within the image is increased

5 Conclusions and Future Work

In this paper, an algorithm for extracting a curve-skeleton from data sets given as point clouds or volumetric data sets was presented. The described algorithm is based on a topological analysis of a vector field derived from the configuration of the point set describing the object's boundary contained in the data set. Due to the fact that it is no longer necessary to compute the vector field on a multitude of points but instead only for points on the object's boundary the described algorithm is significantly faster while still preserving a high accuracy of the extracted curve-skeleton. It took the algorithm a few seconds to extract the curve-skeleton for the smaller data sets. For the porcine heart data set, the algorithm needed a little more than an hour to determine the curve-skeleton. This is comparably fast, considering that in our tests Cornea's et al. [12] algorithm would have required several months to compute the potential field alone.

It is planned to use the described algorithm for deriving precise quantitative measurements from CT scanned specimens, such as vascular structures. In order to be able to measure vessel lengths, vessel diameters, and bifurcation angles an accurate representation of the center lines of the vessels is required. These center lines can be determined by the described algorithm for extracting curve-skeletons applied to the volumetric data set generated by a CT scanner.

Acknowledgments

The author would like to thank Ghassan Kassab and Jenny Choy-Zorrilla for providing the porcine heart data sets, including the manual measurements. I would also like to thank the Wright State University and the Ohio Board of Regents for supporting this research as well as Hang Si and Nicu Cornea for making their source code publicly available and providing data sets through their web pages.

References

1. N. Ahuja, J.-H. Chuang, *Shape Representation Using a Generalized Potential Field Model.* IEEE Trans. Pattern Analysis and Machine Intelligence, 19(2):169–176, 1997.
2. N. Amenta, S. Choi, R.-K. Kolluri, *The Power Crust,* Proc. of 6th ACM Symp. on Solid Modeling, pp. 249–260, 2001.
3. G. Bertrand and Z. Aktouf, *A three-dimensional thinning algorithm using subfields,* Vision Geometry III, 2356:113–124. SPIE, 1994.
4. I. Bitter, A. E. Kaufman, M. Sato, *Penalized-Distance Volumetric Skeleton Algorithm,* IEEE Trans. Visualization and Comp. Graphics, 7(3), 2001.
5. G. Borgefors, *On Digital Distance Transforms in Three Dimensions,* Computer Vision and Image Understanding 64(3):368–376, 1996.
6. G. Borgefors, I. Nyström, G. S. Di Baja, *Computing skeletons in three dimensions,* Pattern Recognition, 32(7), 1999.
7. S. Bouix, K. Siddiqi, *Divergence-Based Medial Surfaces,* ECCV 1842:603–618, Springer-Verlag, 2000.
8. D. Brunner, G. Brunnett, *Mesh Segmentation Using the Object Skeleton Graph,* Proc. IASTED International Conf. on Computer Graphics and Imaging, 48–55, ACTA Press 2004.
9. J. F. Canny, *A Computational Approach to Edge Detection.* IEEE Trans. Pattern Analysis and Machine Intelligence, Vol. PAMI-8, No. 6, pp. 679–698, 1986.
10. M. Couprie and R. Zrour, *Discrete Bisector Function and Euclidean Skeleton,* Lecture Notes in Computer Science, vol. 3429, Springer-Verlag, 2005.
11. N. D. Cornea, D. Silver, P. Min, *Curve-Skeleton Applications.* In Proceedings IEEE Visualization, pp. 95–102, 2005.
12. N. D. Cornea, D. Silver, X. Yuan, R. Balasubramanian, *Computing Hierarchical Curve-Skeletons of 3D Objects.* The Visual Computer 21(11):945–955, Springer-Verlag, 2005.
13. T. K. Dey and S. Goswami. *Tight Cocone: A water-tight surface reconstructor.* Proc. 8th ACM Sympos. Solid Modeling Applications, 127–134. Journal version in J. of Computing and Infor. Sci. Engin. Vol. 30, 2003, 302–307.
14. U. Eckhardt, G. Maderlechner, *Invariant Thinning,* Pattern Recognition and Artificial Intellicgence (7):1115–1144, 1993.
15. N. Gagvani and D. Silver, *Parameter Controlled Volume Thinning,* Graphical Models and Image Processing, 61(3):149–164, 1999.
16. N. Gagvani and D. Silver, *Animating volumetric models,* Academic Press Professional 63(6):443–458, 2001.
17. P. Golland, W. E. L. Grimson, *Fixed Topology Skeletons,* IEEE CVPR, 2000.
18. W. Gong and G. Bertrand, *A simple parallel 3d thinning algorithm.* Proc. IEEE Pattern Recognition, 188–190, 1990.
19. T. He, L. Hong, D. Chen, Z. Liang, *Reliable Path for Virtual Endoscopy*: Ensuring Complete Examination of Human Organs, IEEE Trans. Visualization and Comp. Graphics, 7(4):333–342, 2001.
20. M. W. Hirsch, S. Smale, *Differential Equations, Dynamical Systems and Linear Algebra,* Academic Press, 1974.
21. R. Jain, R. Kasturi, B. G. Schunck. *Machine Vision.* McGraw-Hill,Inc., New York, 1995.
22. A. Kanitsar, D. Fleischmann, R. Wegenkittl, P. Felkel, E. Gröller, *CPR: Curved Planar Reformation.* Proc. IEEE Visualization, pp. 37–44, 2002.

23. T. Lee and R. L. Kashyap, *Building skeleton models via 3-d medial surface/axis thinning algorithms.* CVGIP: Graphical Models and Image Processing, 56(6):462–478, November 1994.
24. S. Lobregt and P. W. Verbeek and F. C. A. Groen, *Three-Dimensional Skeletonization: Principle and Algorithm*, IEEE Transactions on Pattern Analysis and Machine Intelligence, 2(1): 75–77, 1980.
25. C. Lohou and G. Bertrand, *A 3D 12-subiteration thinning algorithm based on P-simple points*, Discrete Applied Mathematics 139:171–195, Elsevier, 2004.
26. V. Luboz, X. Wu, K. Krissian, C. F. Westin, R. Kikinis, S. Cotin, S. Dawson, *A segmentation and reconstruction technique for 3D vascular structures*, MICCAI 2005, Lecture Notes in Computer Science 3749:43–50, 2005.
27. C. M. Ma and M. Sonka, *A fully parallel 3d thinning algorithm and its applications.* Computer Vision and Image Understanding, 64(3):420–433, 1996.
28. G. Malandain, S. Fernandez-Vidal, *Euclidean Skeletons, Image and Vision Computing*, vol. 16:317–327, 1998.
29. A. Manzanera, T. Bernard, F. Preteux, B. Longuet, *A unified mathematical framework for a compact and fully parallel n-D skeletonization procedure*, Vision Geometry VIII, Vol. 3811: 57–68, SPIE, 1999.
30. K. Palagyi, A. Kuba, *Directional 3D Thinning using 8 Subiterations,* Proc. Discrete Geometry for Computer Imagery, Lecture Notes in Computer Science 1568:325–336, 1999.
31. K. Palagyi and A. Kuba. *A parallel 3d 12-subiteration thinning algorithm.* Graphical Models and Image Proc., 61(4):199–221, 1999.
32. D. Perchet, C. I. Fetita, F. Preteux, *Advanced navigation tools for virtual bronchoscopy*, Proc. SPIE Conf. on Image Processing: Algorithms and Systems III, vol. 5298, 2004.
33. C. Pudney, *Distance-Ordered Homotopic Thinning: A Skeletonization Algorithm for 3D Digital Images*, Computer Vision and Image Understanding, 72(3):404–413, 1998.
34. P. K. Saha, B. B. Chaudhuri, D. Dutta Majumder, *A new shape preserving parallel thinning algorithm for 3d digital images.* Pattern Recognition, 30(12):1939–1955, 1997.
35. H. Schirmacher, M. Zöckler, D. Stalling, H. Hege, *Boundary Surface Shrinking - a Continuous Approach to 3D Center Line Extraction*, Proc. of IMDSP, 25–28, 1998.
36. J. A. Sethian, *Fast Marching Methods*, SIAM Review, 41(2):199–235, 1999.
37. H. Si, *TetGen, A Quality Tetrahedral Mesh Generator and Three-Dimensional Delaunay Triangulator*, WIAS Technical Report No. 9, 2004.
38. H. Sundar, D. Silver, N. Gagvani, S. Dickinson, *Skeleton Based Shape Matching and Retrieval*, Proc. Shape Modeling Int'l, 2003.
39. K. Suresh, *Automating the CAD/CAE Dimensional Reduction Process*, ACM Symp. On Solid Modeling and Applications, 2003.
40. S. Svensson, I. Nystrom, G. Sanniti di Baja, *Curve Skeletonization of Surface-like Objects in 3D Images Guided by Voxel Classification*, Pattern Recognition Letters, 23 (12):1419–1426, 2002.
41. A. Telea, A. Vilanova, *A robust level-set algorithm for centerline extraction*, Eurographics/IEEE Symp. On Data Visualization, pp. 185–194, 2003.
42. Y. F. Tsao and K. S. Fu, *A parallel thinning algorithm for 3d pictures.* Computer Vision, Graphics and Image Proc., 17:315–331, 1981.

43. L. Wade, R. E. Parent, *Automated generation of control skeletons for use in animation*, The Visual Computer 18(2):97–110, 2002.
44. M. Wan, F. Dachille, A. Kaufman, *Distance-Field Based Skeletons for Virtual Navigation*, IEEE Visualization 2001, pp. 239–246, 2001.
45. T. Wischgoll, Gerik Scheuermann, *Detection and Visualization of Planar Closed Streamlines*, IEEE Transactions on Visualization and Computer Graphics, 7(2): 165–172, 2001.
46. T. Wischgoll, *Closed Streamlines in Flow Visualization*, Ph.D. Thesis, Universität Kaiserslautern, Germany, 2002.
47. Z. Yu, C. Bajaj, *A Segmentation-Free Approach for Skeletonization of Gray-Scale Images via Anisotropic Vector Diffusion*, CVPR 2004, pp. 415–420, 2004.
48. Y. Zhou, A. Kaufman, A. W. Toga, *Three-dimensional Skeleton and Centerline Generation Based on an Approximate Minimum Distance Field*, The Visual Computer, 14, pp. 303–314, 1998.
49. Y. Zhou, A. W. Toga, *Efficient skeletonization of volumetric objects*, IEEE Trans. Visualization and Comp. Graphics, 5(3):196–209, 1999.

Printing: Krips bv, Meppel, The Netherlands
Binding: Stürtz, Würzburg, Germany